Leveraging Machine Learning for Operation Assessment

DANIEL EGEL, RYAN ANDREW BROWN, LINDA ROBINSON,
MARY KATE ADGIE, JASMIN LÉVEILLÉ, LUKE J. MATTHEWS

Prepared for the United States Army
Approved for public release; distribution unlimited

RAND ARROYO CENTER

For more information on this publication, visit **www.rand.org/t/RR4196**.

About RAND

The RAND Corporation is a research organization that develops solutions to public policy challenges to help make communities throughout the world safer and more secure, healthier and more prosperous. RAND is nonprofit, nonpartisan, and committed to the public interest. To learn more about RAND, visit www.rand.org.

Research Integrity

Our mission to help improve policy and decisionmaking through research and analysis is enabled through our core values of quality and objectivity and our unwavering commitment to the highest level of integrity and ethical behavior. To help ensure our research and analysis are rigorous, objective, and nonpartisan, we subject our research publications to a robust and exacting quality-assurance process; avoid both the appearance and reality of financial and other conflicts of interest through staff training, project screening, and a policy of mandatory disclosure; and pursue transparency in our research engagements through our commitment to the open publication of our research findings and recommendations, disclosure of the source of funding of published research, and policies to ensure intellectual independence. For more information, visit www.rand.org/about/research-integrity.

RAND's publications do not necessarily reflect the opinions of its research clients and sponsors.

Published by the RAND Corporation, Santa Monica, Calif.
© 2022 RAND Corporation
RAND® is a registered trademark.

Library of Congress Cataloging-in-Publication Data is available for this publication.

ISBN: 978-1-9774-0443-5

Limited Print and Electronic Distribution Rights

This publication and trademark(s) contained herein are protected by law. This representation of RAND intellectual property is provided for noncommercial use only. Unauthorized posting of this publication online is prohibited; linking directly to its webpage on rand.org is encouraged. Permission is required from RAND to reproduce, or reuse in another form, any of its research products for commercial purposes. For information on reprint and reuse permissions, please visit www.rand.org/pubs/permissions.

Preface

This report documents research and analysis conducted as part of a project titled *Exploiting Social Media and Operational Data Streams*, sponsored by U.S. Army Special Operations Command. In a previous project for U.S. Army Special Operations Command, which developed a standardized approach for operation assessment of special operations forces, we discovered that a wide variety of data already collected by operational-level headquarters are relevant for assessment but are rarely used. The purpose of this project was to describe and illustrate how machine learning analytics can be used to leverage these existing data to support assessments at the operational level of war. This research is focused on assessment of missions conducted by special operations forces, though the insights in this report are anticipated to be of interest to the broader set of analysts assessing development, humanitarian assistance, and military missions.

This research was conducted within RAND Arroyo Center's Strategy, Doctrine, and Resources Program. RAND Arroyo Center, part of the RAND Corporation, is a federally funded research and development center (FFRDC) sponsored by the United States Army.

RAND operates under a "Federal-Wide Assurance" (FWA00003425) and complies with the *Code of Federal Regulations for the Protection of Human Subjects Under United States Law* (45 CFR 46), also known as "the Common Rule," as well as with the implementation guidance set forth in DoD Instruction 3216.02. As applicable, this compliance includes reviews and approvals by RAND's Institutional Review Board (the Human Subjects Protection Committee) and by the U.S. Army. The views of sources utilized in this study are solely their own and do not represent the official policy or position of DoD or the U.S. Government.

Contents

Figures and Tables

Figures

Tables

Summary

The research reported here was completed in September 2019, followed by security review by the sponsor and the Office of the Chief of Public Affairs, with final sign-off in September 2021.

Accurate and timely assessment—which provides feedback on what is working, what is not, and how operations might be improved—is a critical requirement for all U.S. military operations. Assessments of military operations are structured around the campaign's operational objectives, with the assessment process focused on determining whether progress toward achieving those objectives is being made. The assessment team—ideally in conjunction with the command group—identifies the desired effects associated with each objective and then develops measures of effectiveness (MOEs) to track progress toward achieving these objectives.

Obtaining and processing the right types of data is critical in producing assessments that are relevant and defensible, and this is the subject of this report. Valid assessments require data that are objective, longitudinal, relevant to the types of effects that military operations are hoping to achieve, appropriately structured for assessments, and available with sufficient temporal frequency to support decisionmaking.

In previous research, which developed a standardized approach for operation assessment of special operations forces (SOF), we discovered that a wide variety of data already collected by operational-level headquarters are relevant for assessment but are rarely used. Textual descriptions of unfolding events already available in three types of data—intelligence reporting (such as from human intelligence or signals intelligence sources), operational reporting (such as situation reports [SITREPs]), and ambient data (such as social media)—are often the best available source of information about the local population and enemy and partner forces. However, these data are rarely structured in a way that can be easily integrated into the assessment process.

This report describes one approach for how machine learning (ML) tools can be leveraged to incorporate these existing data into operation assessment. We demonstrate how an ML-based text classifier can rapidly sort through and prepare these data for subsequent analysis using standard statistical tools commonly used by assessment teams. We then illustrate the approach using data from the recently concluded campaign against the Lord's Resistance Army. This ML-based approach should be feasible for most assessment teams and can be implemented with publicly and freely available

ML tools that are pre-authorized for use on U.S. Department of Defense (DoD) classified systems.

The approach described in this report allows the assessment team to derive accurate and timely insights about a campaign, providing the commander with near-real-time results that are objective and statistically relevant. Each type of data that can be analyzed using our approach—intelligence, operational, and ambient—provides a different lens for understanding a campaign's effects. In military operations where assessment-specific data (e.g., polling) are available, these new data sources enable the assessment team to triangulate findings, enhancing the reliability of the assessment. However, ML tools are particularly beneficial in campaigns with limited or no assessment-specific data—which is common in campaigns with limited resources or in denied areas.

Supervised Machine Learning for Operation Assessment

Our analysis focuses on a specific application of ML for assessment: the use of "supervised machine learning" (SML) to construct an assessment-ready database from the unstructured text currently available in existing intelligence, operational, and ambient reporting. The intuition for this application of SML is relatively straightforward. First, human analysts review and analyze, by hand, a subset of the available unstructured text. The output from this first step is typically referred to as "training data." The ML algorithms then analyze this training data and try to mimic the same analytical process used by the human analysts for all remaining text.

This approach allows the assessment team to systematically codify the information in the unstructured text —with the qualitative description provided in each report becoming a single data point in an objective and quantitative assessment—and then examine how this information evolves over time. This approach is the simplest way of incorporating ML into the assessment process, yet it is still an effective approach for deriving assessment-relevant insights from the information-rich unstructured text in these existing data.

In this report, we detail and then illustrate a five-step workflow, summarized in Figure S.1, that assessment teams can follow for incorporating SML into the assessment process. Step 1 is the curation of the large volume of potentially relevant reports—typically thousands of annual reports (or more)—available for most operations. This curation process involves the removal of unnecessary duplicates and the reorganization of data into a format appropriate for analysis by ML algorithms. In our approach, the unstructured text in available reporting is divided into single sentences, which are then analyzed for their relevance to campaign's MOEs.

Step 2 is the development of coding criteria that are appropriate for each relevant MOE. These coding criteria are developed inductively, with the assessment team first reviewing the available data to determine which MOEs are measurable using the avail-

Figure S.1
Supervised Machine Learning for Assessment Workflow

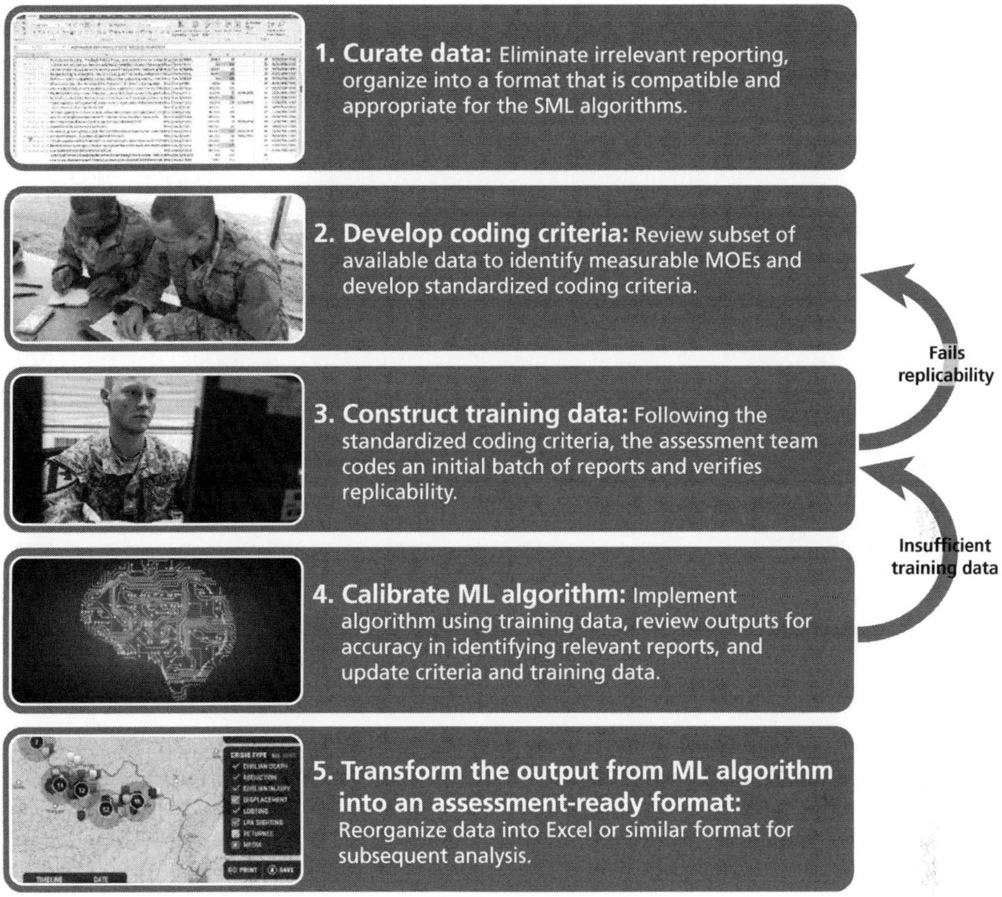

able data. The team then develops specific coding criteria—namely, the set of rules for when a report ought to be considered relevant (or not) for that MOE. It is critical that these coding criteria are sufficiently clear and detailed to allow replicability so that different analysts would produce nearly identical results when applying the criteria. This report provides sample coding criteria for commonly used MOEs for each class of data (intelligence, operational, and ambient), though the specific coding criteria are necessarily campaign-specific.

Step 3 is the preparation of training data, which is the primary input for the ML algorithms in the SML approach. These training data are a subset of the overall unstructured data that are reviewed, analyzed, and coded by the assessment team using the coding criteria developed in Step 2. Ensuring the robustness and replicability of the coding criteria is a key component of the development of these training data, as the ML algorithms seek to replicate the coding of human analysts.

Step 4 is the implementation and calibration of one or a few selected ML algorithms. The calibration process involves reviewing reports coded by the ML algorithms to verify their accuracy—that is, to ensure that reports identified as relevant are indeed relevant. If the algorithms perform poorly, the assessment team needs to adjust algorithm parameters or expand the amount of training data. During an actual campaign, this calibration process should be repeated periodically to ensure that the algorithms still perform well as additional data become available.

In Step 5, the assessment team processes the output from the ML algorithms for integration into the assessment process. The output from the ML algorithms is a structured database (e.g., Excel) that identifies all reports that satisfy the coding criteria developed in Step 2. This common and structured format allows analysis by tools commonly used by assessment teams and facilitates insertion into existing situational awareness tools. In some cases, these data can be analyzed directly to produce trend lines of reporting over time—for example, the percentage of social media reports that discuss an entity (such as an enemy or partner force) in a positive way compared with the number that discuss it in a negative way. In other cases, these data need to be combined with other information available from the report—for example, extracting location-specific information can enable the assessment team to generate evolving maps of enemy freedom of movement.

Recommendations

Our research identified five recommendations that would enhance the value of ML-based assessment approaches:

Recommendation 1: Validate the SML approach in a controlled exercise. The approach described in this report was deliberately designed to be as easy as possible for an assessment team to replicate, and the approach was designed to rely on tools that are both free and already approved for use on many military networks. However, though we have tested our approach with an actual operation, we did not attempt to do this in real time or with a team of military analysts. Thus, we recommend validating the requirements—such as premission training, staffing, availability of analytical tools, and necessary data access—using a live exercise, but under controlled conditions.

Recommendation 2: Explore how unsupervised ML can be used to inform operation assessment. Unsupervised ML can provide a tool for discovering unanticipated patterns between and among a campaign's activities and the operational objectives of those activities. After review by an assessment team, such information can provide the commander a valuable tool with which to explore unanticipated patterns. Given our mandate to explore how ML could be used to leverage existing data for assessment, we did not examine in detail either the practicality or the value of integrating unsupervised ML into assessment.

Recommendation 3: Implement modest standardization of operational reporting. SITREPs and other operational reporting are often the single best historical data source on partner force activities and capabilities, and such reporting can also provide a detailed historical record of the activities of U.S. forces. These characteristics make operational reporting a potentially powerful tool for the assessment team—particularly if ML (or analogous) tools are available to rapidly extract assessment-relevant information—but this reporting is inconsistent in quality and detail. We recommend a modest standardization of this operational reporting: Require (simply) that tactical elements discuss their partner force's activities and capabilities on an ongoing basis.

Recommendation 4: Improve archiving, discovery, and extraction of historical intelligence and operational reporting. Accessing and extracting the relevant intelligence and operational reporting necessary for the approaches described in this report is not easy with existing systems. For intelligence reporting, we recommend that existing systems be augmented with a capability to rapidly extract longitudinal data (e.g., three or more months of data) for a given set of search parameters (e.g., geographic, Boolean logic) that would contain the full text of the report alongside some basic descriptive information about the data (date, source, etc.). For operational reporting, we recommend that operational-level headquarters ensure that SITREPs from all subordinate headquarters are systematically archived.

Recommendation 5: Expand assessment-specific discussions required in professional military education. Despite the acknowledged importance of assessment to military operations, there are major gaps in the assessment-related education and training provided to the joint force. As a consequence, assessment teams often struggle to provide the information necessary to guide effective decisionmaking, and other staff at the operational-level headquarters often lack the experience and training to effectively integrate assessments into operations and planning. The increased complexity of assessments leveraging ML methods is likely to exacerbate these difficulties for headquarters using these methods. We recommend expanding the assessment-related material offered to both officers and enlisted personnel in professional military education, to enable operational-level headquarters to more effectively integrate ML and other emerging technologies into assessment and, more broadly, to prepare the joint force for the gradually increasing assessment requirements for military operations.

Acknowledgements

We are indebted to the many representatives of the U.S. Army special operations community who supported this effort, though we are particularly thankful for insights from the soldiers and civilians at 3rd Special Forces Group (Airborne), USASOC G2, and USASOC G9. We benefited enormously from detailed peer reviews by Elizabeth Hastings Roer of RAND and Jonathan Schroden of the Center for Naval Analyses. We also extend thanks to our RAND colleague Geoff Grimm for his technical support on this project.

Abbreviations

ACLED	Armed Conflict Location & Event Data Project
DoD	U.S. Department of Defense
GEOINT	geospatial intelligence
HUMINT	human intelligence
IIR	intelligence information report
LRA	Lord's Resistance Army
ML	machine learning
MOE	measure of effectiveness
NLP	natural language processing
SIGINT	signals intelligence
SITREP	situation report
SOF	special operations forces
SML	supervised machine learning
SVM	support vector machine
TTPs	tactics, techniques, and procedures
USASOC	U.S. Army Special Operations Command

CHAPTER ONE

Introduction

Accurate and timely assessment at the operational-level of war remains a critical requirement across all U.S. military operations. Operation assessment—which provides "feedback as to what is working, what is not working, and how the force can do things better"[1]—can "make operations more effective" by facilitating "unity of effort and purpose" and allowing the commander and his staff to "proactively identify and adjust to emerging opportunities and risks to mission accomplishment."[2] Yet, although the value of assessment is well understood within the U.S. military,[3] approaches that produce statistically relevant and defensible results have proven elusive.[4]

Obtaining and processing the right types of data is critical in producing assessments that are relevant and defensible, and this is the subject of this report. Valid assessments require data that are objective, longitudinal (i.e., systematically collected over the duration of operations), relevant to the types of effects that military operations are hoping to achieve, appropriately structured for assessments, and available with sufficient temporal frequency to support decisionmaking.

In a previous report, which developed a standardized approach for operation assessment of special operations forces (SOF), we discovered that a wide variety of data already collected by operational-level headquarters are relevant for assessment, but rarely used.[5] This includes intelligence reporting (e.g., human intelligence [HUMINT]

---

[1] Army Doctrine Reference Publication 5-0, *The Operations Process*, Washington, D.C.: Headquarters, Department of the Army, 2012, p. 2-2.

[2] Air Land Sea Application Center, *Operations Assessment: Multi-Service Tactics, Techniques, and Procedures for Operation Assessment*, ATP 5-0.3, MCRP 5-1C, NTTP 5-01.3, AFTTP 3-2.87, 2015, p. ix.

[3] For example, Francis G. Hoffman, "Strategic Assessment and Adaptation: Reassessing the Afghanistan Surge Decision," *Naval War College Review*, Vol. 69, No. 3, 2016, p. 2.

[4] For example, Ben Connable, *Embracing the Fog of War: Assessment and Metrics in Counterinsurgency*, Santa Monica, Calif.: RAND Corporation, MG-1086-DOD, 2012.

[5] Linda Robinson, Daniel Egel, and Ryan Andrew Brown, *Measuring the Effectiveness of Special Operations*, Santa Monica, Calif.: RAND Corporation, RR-2504-A, 2019.

and signals intelligence [SIGINT]),[6] operational reporting (e.g., situational reports [SITREPs]), and ambient data (e.g., satellite imagery, social media). Although they are often the best available source of information about the enemy, local population, and our partner forces, these data are rarely used to support assessments in large part because they are not available in a structured format easily amenable to analysis.

This report augments that previous research by exploring how machine learning (ML) can be used to rapidly extract and prepare these existing data to support operation assessment. We demonstrate how free and publicly available ML tools can rapidly ingest and interpret the large quantities of unstructured text available in intelligence, operational, and ambient data to support analysis that is objective and statistically relevant.[7]

This report should be of particular interest to the U.S. Army and broader U.S. Department of Defense (DoD) special operations community, as this report—in conjunction with a previous report prepared for the U.S. Army Special Operations Command (USASOC)[8]—provides a standardized approach for using existing data to conduct operation assessment of SOF. We believe that the approach that we outline could also be both useful and adaptable to a broad range of missions conducted by the Joint Force.

Operation Assessment: Purpose and Challenges

In current joint doctrine, *operation assessment* is defined as "a continuous activity that supports decisionmaking by ascertaining progress toward accomplishing a task, creating an effect, achieving an objective, or attaining an end state for the purpose of developing, adapting, and refining plans and for making campaigns and operations more effective."[9] Assessment—when integrated into operational planning and the commander's decisionmaking cycle[10]—supports ongoing decisionmaking on the appropriate disposition of forces, needed orders, and plan development by providing

[6] Intelligence and assessment personnel typically work closely together (e.g., Dave LaRivee, "Best Practices Guide for Conducting Assessments in Counterinsurgencies," *Small Wars Journal*, 2011), though the use of actual intelligence reporting in the assessment process is still rare.

[7] Implementations of many widely used algorithms are available in free, open-source software packages. Example ML toolboxes include scikit-learn (Python), Apache Mahout (Java), Libsvm and Liblinear, PyTorch, TensorFlow, and CNTK.

[8] Robinson, Egel, and Brown, 2019.

[9] Joint Publication 5-0, *Joint Planning*, Washington, D.C.: Joint Chiefs of Staff, June 16, 2017, p. VI-1.

[10] U.S. Army War College, *Campaign Planning Handbook*, Carlisle Barracks, Pa.: Department of Military Strategy, Planning, and Operations, 2016; Air Land Sea Application Center, 2015. For SOF-specific guidance on assessment and its relationship to operational decisionmaking, see Gittipong Paruchabutr, Christopher Parapone, Geoffrey Demarest, Grant Martin, Christopher Kirk, Cody Brown, Curtis Clayton, William Harris, Thomas

feedback on what is working (and what is not).[11] Assessments are also often required to satisfy information requirements of national-level authorities (e.g., combatant commands, Congress).

Assessments of military operations are structured around the campaign's operational objectives, with the assessment process focused on determining whether progress toward achieving those objectives is being made.[12] The assessment team—ideally in conjunction with the command group—identifies the desired effects associated with each objective and then develops measures of effectiveness (MOEs) to track progress toward achieving these objectives.[13] These MOEs must be *relevant* (i.e., have a direct link with the objective), *measurable* (with available data), *malleable* (i.e., be reasonably expected to change within the time frame of analysis), and *distinct* (i.e., be mutually exclusive).[14]

Effective assessments of military operations must overcome several core challenges. Foremost among these is *establishing a plausibly causal relationship* between the activities of the military operation and the MOEs being used to measure progress. While this is a general challenge facing assessment,[15] it is a particular challenge for assessments of military operations.[16] *Measurement*—specifically, the difficulty in deriv-

---

Heilman, and Brian Kaiser, *SOF Campaign Planner's Handbook of Operational Art and Design*, Fort Bragg, N.C.: U.S. Army Special Operations Command, September 16, 2014.

[11] ADRP 5-0, 2012, p. 2-2.

[12] Some analysts recommend dividing operational objectives into lines of effort with separate lines of analysis for each (e.g., Alejandro S. Hernandez, Julian Ouellet, and Christopher J. Nannini, "Circular Logic and Constant Progress: IW Assessments in Afghanistan," in Leo J. Blanken, Hy Rothstein, and Jason J. Lepore, eds., *Assessing War: The Challenge of Measuring Success and Failure*, Washington, D.C.: Georgetown University Press, 2015, p. 227), though others caution against this approach (e.g., Jonathan Schroden, Rebecca Thomasson, Randy Foster, Mark Lukens, and Richard Bell, "A New Paradigm for Assessment in Counterinsurgency," *Military Operations Research*, Vol. 18, No. 3, 2013, pp. 5–20). Such critiques emphasize the need to illustrate both (1) how measures of effectiveness (MOEs) derived from lines of effort measure progress toward those lines of effort and (2) how lines of effort work together to achieve campaign objectives.

[13] The *Commander's Handbook for Assessment Planning and Execution* provides an eight-step process for selecting MOEs (Joint Staff, 2011, p. III-9–III-13). These MOEs are often contrasted with measures of performance (MOPs) that measure whether subordinate units are performing well rather than measuring whether the activities of those units are helping accomplish the operational objectives.

[14] These criteria are based on Joint Doctrine Note 1-15, *Operation Assessment*, Washington, D.C.: Joint Chiefs of Staff, 2015, pp. II-8 and A-6.

[15] For a discussion, see David Freedman, "From Association to Causation: Some Remarks on the History of Statistics," *Statistical Science*, Vol. 14, No. 3, August 1999.

[16] This is in part because of the variety of friendly and enemy actors also trying to influence similar outcomes—and in a worst case may "devolve into simply gauging the environment, with little understanding as to how changes in that environment are the result of military activities" (Leo J. Blanken and Jason J. Lepore, "Principals, Agents, and Assessment," in Leo J. Blanken, Hy Rothstein, and Jason J. Lepore, eds., *Assessing War: The Challenge of Measuring Success and Failure*, Washington, D.C.: Georgetown University Press, 2015, p. 8)—and the fact that military operations are frequently trying to influence "human behavior, attitudes, and perception" (Deployable

ing MOEs that can be measured using existing data[17]—is a challenge that often leads to simplified quantification of qualitative data, which has been described as "junk arithmetic."[18] The tension between the time needed for developing rigorous assessment products and the commands' requirement for rapid insights to guide planning and execution of operations,[19] which might be described as a challenge of *timeliness,* is anticipated to grow increasingly difficult to mitigate.[20] And a fourth core challenge is the difficulty in *maintaining objectivity*, as both assessment teams and operational units—if they are knowingly integrated into the assessment process—will be incentivized (whether consciously or not) to provide optimistic assessments.[21] This last chal-

---

Training Division, Insights and Best Practices Focus Paper: Design and Planning, Suffolk, Va.: Joint Staff J7, 2013, p. 1).

[17] Assessments of military operations frequently employ available data as a proxy for the effect they want to measure, which can create misleading results (e.g., Daniel Egel and Peter Glick, *Assessing the Stabilizing Effects of Development Programming: A Review of USAID's Evaluation Efforts in Afghanistan*, Santa Monica, Calif.: RAND Corporation, WR-1106-MSI, 2016). Recognizing this challenge, another recommendation is to "not ignore data that are hard, but necessary, to measure. Report them early in the planning process to the commander. Not allocating collection resources to necessary measures puts the assessment quality at risk, and the commander needs to decide whether to accept that risk, reallocate resources to collection, or adjust the assessment plan" (Air Land Sea Application Center, 2015, p. 21).

[18] For a discussion of this challenge and ways to rigorously convert qualitative to quantitative data in military assessments, see William P. Upshur, Jonathan W. Roginski, and David J. Kilcullen, "Recognizing Systems in Afghanistan: Lessons Learned and New Approaches to Operational Assessments," *Prism*, Vol. 3, No. 3, June 2012.

[19] Successful integration of assessment into the operations process gives commanders a means of proactively identifying and adjusting to emerging opportunities and risks to mission accomplishment. Timely recognition of opportunities and risks affords commanders a distinct advantage, possibly allowing them to catch the enemy off-balance and rapidly end a battle; refocus joint force capabilities to minimize disruption; or hasten accomplishment of objectives, conditions, and end states. Conversely, missed opportunities and risks result in protracted engagements, higher casualties, increased potential for tactical defeat, and operational and strategic setback. The challenge of timeliness has been identified as a "contributing factor" in what has been described as an overall lack of success of assessments in recent military operations, with the "few documented successes . . . achieved only after much effort and revision" (Air Land Sea Application Center, 2015, p. ix).

[20] Mushen and Schroden highlight that this challenge of "meeting expectations of modern audiences without compromising the integrity of the assessment may prove even more challenging in the future" (Emily Mushen and Jonathan Schroden, *Are We Winning? A Brief History of Military Operations Assessment*, Washington, D.C.: CNA, September 2014 , p. 37).

[21] As an example, operational headquarters are typically incentivized to produce overly optimistic assessments for external audiences:

> Assessments that reflect bottom-up data collection and analysis, while ostensibly objective, risk creating dissatisfaction among higher leadership if the results do not show sufficient progress. As a consequence, unlike the skeptical approach commanders often take to assessments in the planning process, those delivered "up and out" to policymakers often reflect an optimism that is either not present in or even counter to internal, bottom-up assessments. (Mushen and Schroden, 2014, p. 36)

lenge can be particularly insidious, as it may induce operational units to engage in undesirable behavior.[22]

Incorporating Machine Learning into Operation Assessment

Assessments have typically relied on two classes of data. The first is data that are easily quantified—such as frequency and location of enemy engagements, survey data of local nationals, and satellite imagery of ambient light at night. The second class of data is qualitative assessments, often self-assessments, of progress toward achieving specified campaign objectives (e.g., performance of a partnered force). Both types of data can provide valuable insights to support the assessment process, but both are well known to provide an incomplete (or worse, biased) understanding of evolving conditions on the ground and can be resource-intensive to collect in terms of personnel, time, and funding.

However, there is a broad range of data available for military operations that are relevant for assessments, but rarely used for this purpose. This includes intelligence reporting, which describe the activities and disposition of enemy forces; operational reporting, which describe the activities of U.S. and (where applicable) partner forces; and ambient data (such as social media, traditional media, and satellite imagery), which can be used to assess economic conditions, public opinion, and other population-focused outcomes. These data are often the best available source of information about the enemy, local population, and our partner forces, but they are rarely—if ever—used to support assessments. The three central challenges in using these data for assessment is that they are (1) not perceived as sufficiently objective, (2) not available in a structured format easily amenable to analysis, and (3) extremely numerous and require some effort to obtain and organize.

In this report, we describe and illustrate how an assessment team can leverage publicly available ML tools to integrate these existing into the assessment process. These tools allow the assessment team to rapidly ingest and interpret large quantities of unstructured text—from intelligence reporting, operational reporting, and social media—to support analysis that is objective and statistically relevant. We also show how these tools can be automated to provide near-real-time updates on unfolding conditions on the ground. Critically, the core value of these tools is in making new data and insights available to the team—while they may provide a more efficient way of

[22] For example, Robert B. Sotire, *Measuring Performance and Effectiveness in Irregular Warfare: Preventing Dysfunctional Behavior*, Newport, R.I.: Naval War College, 2009.

producing some of the quantitative components of an assessment, they do not supplant the critical qualitative elements.[23]

The intuition for our application of ML is straightforward.[24] In a best possible world—at least from the assessment team's perspective—commanders would task their forces to systematically and objectively report information that would allow the commander to track progress toward campaign objectives. If forces can effectively and efficiently provide this type of feedback, then the approach presented in this report offers limited value to the assessment team.[25] However, there will be many cases in which assessment-specific reporting is not available or where its objectivity is likely to be questioned. In these cases, the ML tools presented in this report can provide an objective and systematic "estimate" of how these individuals would have responded to the commander's assessment-specific information requirements by rapidly reviewing the reporting already generated by operational units, intelligence analysts, and the local population (e.g., social media).

Research Approach

Our approach for exploring how ML might be leveraged to support operation assessment—focusing on the specific requirements of operational-level SOF headquarters—relies on three types of research activities. The first is an analysis of how "supervised machine learning" (SML) can be used to extract assessment-relevant insights from these otherwise unstructured data. Our analysis focuses on unstructured text data—the most prevalent form of data currently available—and examines how an SML approach can be used to extract a structured and longitudinal dataset aligned with a campaign's MOEs from existing intelligence, operational, and ambient data.

The intuition behind SML is relatively straightforward, in that the assessment team must first analyze, by hand, a subset of these data and the ML algorithms then mimic the same analytical process across all available data. In the application considered here, the SML algorithm is tasked to infer, for each unstructured text report in a large sample, whether that report is relevant or irrelevant to a specific MOE. Analogous tools can be used to systematically and objectively analyze the audio, imagery, and video currently collected by a multitude of sensors (e.g., cell phones, unmanned

---

23 For a discussion of the importance of qualitative methods, see Jonathan Schroden, "Why Operations Assessments Fail: It's Not Just the Metrics," *Naval War College Review*, Vol. 64, No. 4, Autumn 2011.

24 We thank Jon Schroden for suggesting this analogy.

25 The use of our approach to analyze social media and other ML approaches (e.g., analysis of imagery or video), as discussed in the following paragraph, would still have value.

aerial vehicles, satellites), though we believe that harnessing unstructured text has the most value, by far, because it already represents a degree of interpretation.[26]

The second component explores the specific types of assessment-relevant insights that can be extracted from existing intelligence, operational, and ambient data using this SML-based approach. This analysis is based on a systematic review of the data already collected and available for military operations. This review describes the types of data available, the existing systems available for extracting and analyzing those data, and the types of MOE-specific insights that can be derived from each type of data. Our review was based on a systematic review of data collected by Army SOF, though the data identified in this review are generally available across all U.S. military operations.[27]

The final component is a demonstration of this approach using data from Operation Observant Compass, the recently concluded campaign against the Lord's Resistance Army (LRA) in central Africa. The purpose of this example is not to provide a robust assessment of Operation Observant Compass nor to demonstrate how an assessment team might optimize the SML tools, but rather to illustrate the viability of the SML-based approach and the types of novel, real-time insights that can emerge.

Organization of This Report

Chapter Two describes how SML can be used to extract systematic and objective assessment-relevant insights from unstructured data, providing a detailed step-by-step process to do so. Chapter Three then examines how this approach can be applied to the diversity of existing data available for assessment, illustrating the types of MOE-relevant insights that can be extracted from each. Chapter Four uses the recently concluded campaign against the LRA to illustrate the approach, demonstrating how one specific ML tool can be used to develop a structured dataset that can then be analyzed and visualized using standard assessment approaches. Chapter Five concludes with several recommendations—describing the types of practical exercises, improvements to existing data collection systems, and training that ought to be implemented to maximize the value of SML—and discusses the next steps and risks from using these ML tools. Two appendices describe the specific SML tool that we used in Chapter Four and provide replication code for both implementing this specific SML tool and analyzing the data that emerge from its implementation. A classified appendix (not available in this unclassified version) provides some additional detail for discussions throughout the report.

---

[26] It may also include bias, particularly in the case of operational reporting.

[27] This survey relied on consultations with USASOC and 3rd Special Forces Group (Airborne) personnel, reviews of existing documentation, discussions with other RAND personnel, and the experience of the research team.

CHAPTER TWO

Supervised Machine Learning and Assessment

This chapter describes how ML can be leveraged to support assessment. Our discussion focuses on a specific application of ML for assessment of military operations: the use of "supervised machine learning" (SML) to construct an assessment-ready database from the unstructured text available in intelligence reporting, operational reporting, and ambient data (e.g., social and traditional media). While the bulk of this chapter focuses on this single application, we also discuss broader applications of ML to the assessment process.

The next section provides an overview of this application of SML to assessment. The intent of this section is not to provide a comprehensive review of SML—which is beyond the scope of this report—but instead to describe one approach for integrating SML into the assessment process and the value of doing so. Our approach uses SML to rapidly identify assessment-relevant information in otherwise unstructured reporting and then convert this information into a format appropriate for further analysis using standard assessment tools.

We then provide a step-by-step description of how an assessment team would use SML to analyze unstructured text. The intuition for our approach is relatively straightforward, in that the assessment team first analyzes, by hand, a subset of the unstructured text and the ML algorithms then try to mimic the assessment team's analytical approach for the remaining data. This section details a five-step process that assessment teams can employ to implement this approach, and Chapter Three describes how it can be applied to each of three major data types (intelligence, operational, and ambient).

A final section concludes by offering a broader discussion of how SML might be used for other data formats likely to contain assessment-relevant information (e.g., imagery, video). For each, we discuss the tools that are available, the types of inference that are possible, and the types of resources necessary to extract these insights—at least given currently available technology and tools.

Supervised Machine Learning and Its Application to Assessment

A common application of ML is "knowledge extraction," in which task-relevant information embedded in large volumes of data (e.g., social media, stock market transactions, medical records) is automatically located and converted into a format for further analysis. One ML-based approach for this knowledge extraction is SML, in which human analysts are actively involved in training ML algorithms to recognize patterns in data.[1]

The intuition for this SML approach is relatively straightforward. First, human analysts review and analyze, by hand, a subset of the available data. The output from this first step is typically referred to as "training data." The ML algorithms then analyze this training data and try to mimic the same analytical process used by the human analysts for the remaining data.

In our context, SML can be used to rapidly analyze the large volume of unstructured text data typically available for a military operation and determine whether each report provides information salient to the assessment of that operation. The assessment team first builds training data by reviewing a subset of available reporting (such as intelligence reports) and determining whether each report contains information indicative of a particular output of interest relevant for the assessment (such as the location or capabilities of enemy units). The ML algorithm then analyzes this training data, examining the specific words and phrases used in the text of reports indicated as relevant (or not) by the assessment team, and estimates the probability that each remaining report also contains similar assessment-relevant information.

This application of SML allows an assessment team to rapidly and systematically extract assessment-relevant insights from otherwise unstructured data. For example, the assessment team might review a few hundred intelligence reports to "train" the computer, and then infer outputs in thousands of additional reports. This allows the assessment team to incorporate materials into the assessment process that would have otherwise only been consulted on an ad hoc basis,[2] and has the added advantage of being objective, systematic, and transparent—imperatives for an assessment team.

Another core strength of this approach is that it is relatively easily integrated into the assessment process. This relative ease is perhaps best demonstrated by the increased use of SML in the impact evaluation literature—the civilian development community's analogue to assessment—where small teams of academics and development specialists are now integrating SML into assessment using freely and publicly available

[1] Besides SML, another form of ML likely to be useful for assessment is "unsupervised" ML, where learning is performed on unlabeled data. This is discussed in Chapter Five.

[2] These materials are typically used for to provide background, context, or illustrations of current conditions.

ML tools.[3] As examples, researchers have used SML to analyze satellite imagery data to characterize subnational poverty,[4] agricultural production, and damage to physical infrastructure,[5] and researchers have used SML to analyze qualitative interviews with program beneficiaries.[6]

While incorporating ML algorithms into the assessment process is relatively straightforward (as we illustrate in Chapter Four), any implementation of these approaches requires an assessment team with the capability to effectively calibrate these algorithms. Further, in almost all cases, ML algorithms will need to be periodically retrained as part of a "continuous delivery/deployment" process.[7] An assessment team would therefore need a technical specialist who would be responsible for monitoring the performance of the algorithms,[8] and then either updating the parameters of the algorithm or the training data until a satisfactory level of performance is achieved.[9] It is worth noting that human analysts play a similar role in even the most well-resourced and sophisticated applications of SML—for example, IBM's use of ML to develop tai-

---

[3] These tools are currently being employed by both the U.S. Agency for International Development and the World Bank (David McKenzie, "How Can Machine Learning and Artificial Intelligence Be Used in Development Interventions and Impact Evaluations?" *World Bank Blog*, March 5, 2018; Amy Paul, Craig Jolley, and Aubra Anthony, *Reflecting the Past, Shaping the Future: Making AI Work for International Development*, Washington, D.C.: U.S. Agency for International Development, 2018). Each of these teams appears to include at least one assessment specialist, one data specialist, and one ML specialist, though the other requirements are modest.

[4] Neal Jean, Marshall Burke, Michael Xie, W. Matthew Davis, David B. Lobell, and Stefano Ermon, "Combining Satellite Imagery and Machine Learning to Predict Poverty," *Science*, Vol. 353, No. 6301, August 19, 2016.

[5] Andre Gröger, Jonathan Hersh, Andrea Matranga, and Hannes Müller, "Analyzing Conflict from Space: Identification of Physical Destruction During the Syrian Civil War," unpublished research, 2018.

[6] The specific application that we are familiar with relies on unsupervised ML techniques for thematic analysis, rather than the identifying specific outcomes associated with the interview data (Ramya Parthasarathy, Vijayendra Rao, and Nethra Palaniswamy, "Deliberative Inequality: A *Text-as-Data* Study of Tamil Nadu's Village Assemblies," *World Bank*, Policy Research Working Paper 8119, June 2017).

[7] The need for continuous delivery/deployment is emphasized, for example, in DoD's Project Maven (Loren Blinde, "DoD Announces 2nd AI Industry Day," *Intelligence Community News*, October 19, 2018).

[8] Monitoring methods would typically include either cross-validation or a bootstrap estimator to compute performance metrics (Simone Borra and Agostino Di Ciacco, "Measuring the Prediction Error: A Comparison of Cross-Validation, Bootstrap and Covariance Penalty Methods," *Computational Statistics and Data Analysis*, Vol. 54, pp. 2976–2989.)

[9] In some cases, it might be necessary to change the SML algorithm being used. Note that performance metrics and an associated confidence interval may be reported by the assessment team when referencing results of the analysis (Ron Kohavi, "A Study of Cross-Validation and Bootstrap for Accuracy Estimation and Model Selection," *International Joint Conference on Artificial Intelligence*, 1995).

lored recommendations for cancer treatments,[10] or the finance industry, where ML has replaced entire classes of traders.[11]

Step-by-Step Process for Implementing Supervised Machine Learning for Assessment

This section describes a step-by-step process through which an assessment team can use SML to transform otherwise unstructured data into data that can be immediately integrated into the assessment process. The focus of this section is on data available in an unstructured or semistructured text format, with some type of metadata (e.g., date-time group, author, locations mentioned in the report) accompanying the text reporting, though SML of other data types would follow a similar workflow,

This workflow has five steps, which are summarized in Figure 2.1 and discussed in detail throughout the rest of this section. Steps 2 through 4 may need to be repeated, as described below, to ensure that the algorithm is performing well.

Step 1: Curate Data

For most operations, there is a large volume (frequently thousands of reports or more per year) of potentially relevant intelligence, operational, and ambient data available. The first major step in the curation process is to filter available reporting to eliminate reports that are obviously irrelevant. As an example, for enemy-centric MOEs, the assessment team would want to remove all reporting that does not mention the enemy force at least once. Similarly, partner force–focused assessment would remove all reporting that does not mention at least one of the major partner forces.

The second step is to remove unnecessary duplicates. Duplication of information is common across all three types of data—for example, intelligence reporting often includes the same text included in an earlier report, and social media is replete with "re-tweets" (or equivalent) duplication of earlier information. The assessment team might want to retain or remove duplicate reports depending on whether they provide useful analytic insights (e.g., index popularity of themes) or are an unwanted product of search processes or repetition in the database and would therefore bias results.[12]

---

[10] "Watson for Oncology" uses an SML approach—relying on human analysts who "laboriously feed Watson information about how patients with specific characteristics should be treated"—to predict treatments based on available doctor's notes, patient vitals, etc. (Casey Ross and Ike Swetlitz, "IBM Pitched Its Watson Supercomputer as a Revolution in Cancer Care. It's Nowhere Close," *STAT*, September 5, 2017).

[11] For example, Evelyn Cheng, "Just 10% of Trading Is Regular Stock Picking, JP Morgan Estimates," *CNBC*, June 13, 2017; Conrad De Aenille, "A.I. Has Arrived in Investing. Humans Are Still Dominating," *New York Times*, January 12, 2018.

[12] Some databases, such as LexisNexis, have built-in duplicate filtering algorithms that allow the user to remove duplicates before downloading data.

Figure 2.1
Supervised Machine Learning for Assessment Workflow

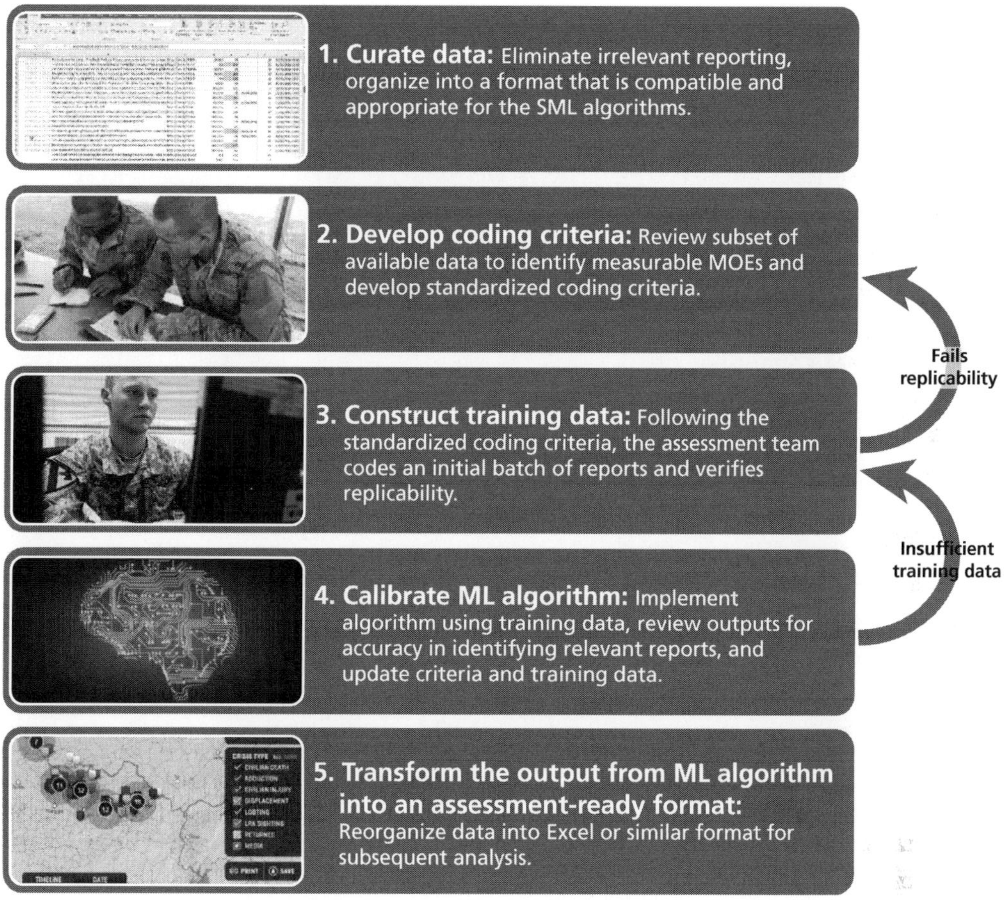

The third step is to reorganize the data in a format that is appropriate for analysis by the ML algorithms. In our approach, the unstructured text of reporting—which could be short excerpts or multipage discussions—is divided into single sentences, and sentences without assessment-relevant keywords (e.g., mention of enemy or partner force) are removed. The analysis then focuses on this subset of sentences, and the neighboring sentences (before and after) because they may provide valuable context and information. Restricting analysis to these sentences may eliminate some potentially relevant reporting, but doing so greatly facilitates the development of coding criteria and training data (Steps 2 and 3) and the performance of the algorithms (Step 4).

Step 2: Develop Coding Criteria

The development of coding criteria, the rubric that the assessment team uses to determine whether the information contained in available data is indicative of a particular output of interest, is an inductive process. The first component of this step is a review of a subset of the data prepared in Step 1 to determine which MOEs are measurable with the available data. The reasoning for this component is straightforward: There is likely to be a range of MOEs that the assessment team would like to include in their analysis, but each available unstructured data source typically provides information on only a subset of these MOEs.

The second is the development of specific coding criteria for each MOE in each relevant data source. These coding criteria are the set of rules that determines whether a report is relevant or not.[13] It is critical that these coding criteria are sufficiently clear and detailed to allow replicability, such that different analysts would produce nearly identical results when applying the criteria. This replicability is particularly important for SML, as the ML algorithms need consistency to be able to effectively mimic the coding of human analysts.

Chapter Three provides some generalized examples of coding criteria for MOEs that are likely to be common across many campaigns. However, development of these coding criteria is necessarily an inductive process, and reviewing available data to prepare campaign-specific coding criteria is a critical step.

Step 3: Construct Training Data

In our context, the assessment team will need to construct training data by hand, reviewing a subset of the overall unstructured data and applying the coding criteria developed in Step 2 to these data. In general, the accuracy of the ML algorithms is higher when there is more training data available, though there is no firm rule for determining the right number of reports to include in the training data. In our experimentation, we have found that training data with approximately 300 reports can be sufficient to train an algorithm at a useful level of accuracy for most MOEs. However, for MOEs that are only infrequently addressed in available reporting, a larger number may need to be reviewed.[14]

Ensuring the replicability, or standardized application, of the coding criteria is a key component of the development of these training data as the SML will use these data to establish patterns that are predictive of relevance. The best way to test replicability is to have multiple members of the assessment team apply these coding criteria

---

[13] The analyst will typically assign a value of "1" (or something equivalent) to indicate that it is relevant, and "0" otherwise. It is this value of "1" that the SML algorithm will be trained to infer.

[14] In some cases, it may make sense to do the SML process twice. The first iteration of the SML would identify a subset of the total reports that are likely to be of relevance and then the second iteration would develop relevant training data on only this restricted sample. As an example, if the MOE of interest was perceptions of the partner force among the local population, it might make sense to do a first iteration of the SML focused on isolating only those reports that mention a partner force and then the second round could focus on identifying reports that mentioned something positive, negative, or neutral.

to the same subset of data and then compare the results from each.[15] As a rough rule of thumb, an interrater reliability of at least 70 percent is necessary for the SML algorithms to perform well.[16] If this threshold is not reached, then the team can confer to iron out any disagreements and refine or clarify the coding criteria as needed.[17]

Step 4: Calibrate Machine Learning Algorithm

Calibration is the iterative process through which the assessment team "fine-tunes" the ML algorithm to produce the highest accuracy in the analysis of the training data.[18] This accuracy is assessed by training the algorithm on a subset of the training data and then applying the trained algorithms to the remaining training data.[19]

This calibration process requires that the assessment team iteratively adjust a number of algorithm parameters. These parameters include the type of information that is given to the algorithm (e.g., single words, clusters of words), the number of unique words or groups of words that are included in the analysis, and some more technical parameters (e.g., kernel function, a regularization parameter that determines the cost of misclassifying samples), which are discussed in more detail in Appendix B. During an actual campaign, this calibration process should be repeated periodically to ensure that algorithms still perform well as additional data become available.

Step 5: Transform the Output from the ML Algorithm into an Assessment-Ready Format

The output from the ML algorithm is a structured database (e.g., Excel) that identifies all reports that satisfy the coding criteria developed in Step 2. In some cases, these data can be analyzed directly to produce trend lines of reporting over time—for example, the percentage of social media reports that discuss an entity (enemy, partner force, host

---

[15] An alternative is to do diagnostics using the ML tool itself, running the ML algorithms on a subset of the training data and then assessing how well it predicts the remaining reports within the training data. The challenge with this approach is that the analyst must supply a single diagnostic tool to validate the quality of the training data (step 3) and the calibration of the algorithm itself (step 4 below).

[16] In practice, interrater reliability can reach 85–90 percent when coding criteria are well articulated for the available data (e.g., Loren Collingwood and John Wilkerson, "Tradeoffs in Accuracy and Efficiency in Supervised Learning Methods," *JITP 2011: The Future of Computational Social Science*, 2011).

[17] See John L. Campbell, Charles Quincy, Jordan Osserman, and Ove K. Pedersen, "Coding In-Depth Semistructured Interviews: Problems of Unitization and Intercoder Reliability and Agreement," *Sociological Methods and Research*, Vol. 42, No. 3, 2013.

[18] We use the term *accuracy* in a generic sense to refer to various measures of performance, such as precision, recall, area under the curve, or the f1-score.

[19] This can be done in a variety of ways, typically employing "cross-validation" or the "bootstrap estimator." Details on these procedures with a recent appraisal of the various techniques can be found in Borra and Di Ciaccio, 2010. Evaluating algorithm performance on data excluded from training reduces the potential risk of "overfitting," whereby the algorithm appears to perform accurately but only because it is evaluated on the same data that it was trained on.

government) in a positive way compared with the number that discuss it in a negative way. In others, these data need to be combined with other information available from the report—for example, extracting location-specific information can enable the assessment team to generate evolving maps of enemy freedom of movement.

In addition to allowing assessment teams to analyze these data using standard analytical approaches, this structured database format facilitates ingestion into existing situational awareness tools. It further ensures transparency and allows these data to be cross-validated against other sources of information about the campaign.

With a limited amount of additional effort, the preparation and analysis of these data can be automated so that the commander can have access to assessments that are constantly being updated based on the most recently available intelligence, operational, and ambient data. This would allow the commander to receive updates as frequently as daily—comparable to the ubiquitous intelligence and operational updates—on progress toward operational objectives. We anticipate that this would stimulate a command-wide discussion that could improve the efficacy of planning and operations, though we are not aware of any assessments that have deployed such an approach.[20]

Other Applications of Supervised Machine Learning to Assessment

This chapter has focused on the use of SML to build assessment-relevant data—namely, data that can be analyzed directly using standard assessment toolkits—from data containing primarily unstructured text. This focus reflects the ubiquity of unstructured text data in military operations, the richness of information contained in this existing reporting, and the comparative ease with which these data can be incorporated into the assessment process. In other words, we have focused on this specific application of ML because it is "cost-effective," at least in relative terms, for the assessment team.

However, there are data in a variety of other formats—for example, various forms of imagery and video—that could be incorporated into operation assessment using an SML approach analogous to that described in this chapter. Leveraging these data for assessment requires somewhat different analytical skills and computational tools, but particularly well-resourced assessment teams might consider doing so. This section describes the strengths and limitations of the existing ML tools for analyzing these data and provides an outline of how assessment-relevant insights might be derived from each.

---

[20] Schroden et al., 2013, observed that assessments "designed to stimulate discussions leading directly to decisions by senior leaders on actions they could take, direct, or request" was a best practice. Though their approach for stimulating this discussion was different, we anticipate that integrating assessment into the daily battle cycle could generate discussion similar to that currently had for intelligence and operational updates.

Imagery

Satellites carry a variety of sensor payloads—including electro-optic cameras, infrared cameras, and synthetic aperture radar—that have been used for tasks such as object detection, change detection, land cover classification, and missile launch detection. Though the classification of insights from these data are often done manually,[21] making it far too resource-intensive to include in assessment, improvements in hardware, algorithms, and data are now making ML-driven analysis of these data possible.

As examples, advanced ML methods are now being developed that will allow analysts to use satellite imagery to systematically assess vehicular traffic (e.g., proxy for economic activity, measure of enemy use of a known bed-down location, etc.),[22] derive detailed assessments of agricultural activity (e.g., proxy for economic activity or measure of production of illicit crops),[23] measure either damage or improvements to infrastructure (e.g., impact of military operations, effectiveness of reconstruction activities),[24] or rapidly detect enemy activity (e.g., surface-to-air missiles).[25] However advanced, these methods are still mostly dependent on hand-coding by human analysts,[26] analogous to the application of SML to unstructured text presented in this chapter. They also face a variety of persistent issues, foremost of these perhaps is that space-based sensors typically do not produce images of sufficient quality (referred to in the literature as "low pixels-on-target") to allow ML algorithms and other automated tools to accurately and consistently identify buildings, roads, vehicles, or other types of human activity.

---

[21] As one example, a 2015 *Satellite Imagery Interpretation Guide* concluded—highlighting the importance of the ML tools that we mention in the next sentence in the text—that the "automation of this identification process is critical for the advancement of humanitarian applications of satellite imagery because, currently, images are manually interpreted by analysts—often a laborious and time consuming process" (Signal Program on Human Security and Technology, *Satellite Imagery Interpretation Guide: Intentional Burning of Tukuls*, Harvard Humanitarian Initiative, 2015).

[22] For example, Chenxi Yang, Wenjing Li, and Zhiyong Lin, "Vehicle Object Detection in Remote Sensing Imagery Based on Multi-Perspective Convolutional Neural Network," *International Journal of Geo-Information*, Vol. 7, No. 7, 2018.

[23] Aaron E. Maxwell, Timothy A. Warner and Fang Fang, "Implementation of Machine-Learning Classification in Remote Sensing: An Applied Review," *International Journal of Remote Sensing*, Vol. 39, No. 9, 2018.

[24] Austin J. Cooner, Yang Shao, and James B. Campbell, "Detection of Urban Damage Using Remote Sensing and Machine Learning Algorithms: Revisiting the 2010 Haiti Earthquake," *Remote Sensing*, Vol. 8, No. 10, 2016.

[25] For example, a method of detecting Chinese surface-to-air missiles that is 81 times faster than the traditional approach: Richard A. Marcum, Curt H. Davis, Grant J. Scott, and Tyler W. Nivin, "Rapid Broad Area Search and Detection of Chinese Surface-to-Air Missile Sites Using Deep Convolutional Neural Networks," *Journal of Applied Remote Sensing*, Vol. 11, No. 4, 2017.

[26] For examples, see Mark Pritt, and Gary Chern, "Satellite Image Classification with Deep Learning," *2017 IEEE Applied Imagery Pattern Recognition Workshop*, 2017, and Fisher Yu, Ari Seff, Yinda Zhang, Shuran Song, Thomas Funkhouser, and Jianxiong Xiao, "LSUN: Construction of a Large-Scale Image Dataset Using Deep Learning with Humans in the Loop," *arXiv*, 2015.

Analogous techniques are being developed to analyze and classify data from the large quantity of images being collected by civilians in the field and available via social media and other platforms. Analysts have used an SML approach to extract insights from Instagram,[27] and the Defense Advanced Research Projects Agency is developing ML algorithms capable of automatically indexing this type of data to facilitate subsequent search.[28] Before this approach can be applied to operation assessment, it is likely that further improvements will be needed to handle the wide variability in viewpoints, camera models, illumination, scenery, and the relatively open-ended nature of the indexing (that is, the fact that multiple objects and/or scene labels need to be extracted from each image).

Video

The analysis of video is another area where ML algorithms are making significant progress, with video-specific algorithms being developed to leverage information present across multiple frames and to perform tasks such as surveillance, event recognition, target tracking, crowd monitoring, and pattern-of-life analysis. From an assessment perspective, these algorithms currently have two types of uses. The first is to support analysis of the full-motion video data being collected by aerial assets—primarily unmanned aerial vehicles that often provide near-continuous coverage for high-value missions—owned by the U.S. military.[29] Like imagery, the low pixels-on-target is likely to remain a limiting factor, though this can be mitigated to some extent with improvements in sensing capabilities and the use of sensor placement capabilities to improve viewpoints.

The second use is the analysis of ground-based video, either from civilian assets (e.g., YouTube) or persistent surveillance networks (e.g., security cameras).[30] The ubiquity of these data—particularly in urban areas—means that there is a large collection of video data potentially available for operation assessment, and they can provide multiple views to increase confidence of findings. ML algorithms have already been developed for tasks such as people detection, crowd counting, and facial recognition, but additional developments are necessary to fully leverage these data.

---

[27] For example, Kay L. O'Halloran, Sabine Tan, Peter Wignell, John A. Bateman, Duc-Son Pham, Michele Grossman, and Andrew Vande Moere, "Interpreting Text and Image Relations in Violent Extremist Discourse: A Mixed Methods Approach for Big Data Analytics," *Terrorism and Political Violence*, Vol. 31, No. 3, 2016.

[28] Defense Advanced Research Projects Agency, "Visual Media Reasoning (VMR) (Archived)," webpage, undated.

[29] Daniel McCready, "Developing Remotely Piloted Air-to-Air Combat Capabilities," Unmanned Systems Summit, Alexandria, Va., March 14, 2019.

[30] A relevant example for civilian assets is the YouTube-8M Video Understanding Challenge that features a large set of labeled YouTube videos used to train ML algorithms to detect actions (YouTube, "YouTube-8M Segments Dataset," webpage, undated). For persistent surveillance, face recognition from surveillance cameras is a leading example (Axis Communications, "Solutions by Application," webpage, undated).

CHAPTER THREE

Operationalizing Available Data for Assessment

There is a diversity of data collected during military operations that could be, but rarely are, integrated into assessments performed by operational-level headquarters. This includes *intelligence data* collected, analyzed, and synthesized across the U.S. government; *operational data* collected by subordinate units; and *ambient data,* including social media activity (e.g., Twitter), satellite-imagery, traditional media (e.g. newspapers), and data collected by other organizations (e.g., civil society organizations). Each of these categories of data provides a different lens for understanding the desired effects of a military operation, but these data are rarely integrated into assessment efforts because they are typically not available in a format that is amenable to standard assessment toolkits. Rather than being available in a structure database format (e.g., incident data that provides the time, place, and intensity of a security-related event), they are available as unstructured text (e.g., text available in a SITREP or newspaper report, voice intercept) or other unstructured data format (e.g., imagery, video).

In this chapter, we describe how the SML approach described in the previous chapter can be used to structure these data so that they can be incorporated into the assessment process. For each data category, we provide a general overview of the types of available data and the relevance of these data to common objectives for military operations. Each of these discussions also includes a few demonstrative examples of how our text-based SML approach can be used to derive assessment-relevant insights.

The following three sections describe, sequentially, this application of SML to intelligence, operational, and ambient data. We then outline a step-by-step process for the assessment team to extract and organize these data in preparation for deployment, and then we briefly review how these data can be integrated into existing operation assessment approaches.

Intelligence Data

Products produced by the intelligence community are typically the best source of enemy-centric data available, providing insights on enemy location, force structure, activities, and evolving enemy intent. In addition, depending on the military operation and the intelligence requirements of the commander, this reporting can also provide insights on the local population, partner forces, and other relevant battlespace conditions.[1]

Three classes of intelligence reporting are the most useful to assessment. The first is human intelligence (HUMINT), which is the set of observations about the battlespace provided by individuals—whether HUMINT collectors, who receive information from informants and other sources, or others with relevant knowledge and insights (such as travelers or refugees).[2] The second is signals intelligence (SIGINT), which includes communications and other transmissions intercepted via electronic means.[3] The third is geospatial intelligence (GEOINT), which consists of photographic and other imagery intelligence fused with geographic and geo-referenced information (e.g., population data from census or other sources).[4] For some specialized operations (e.g., countering weapons of mass destruction), insights from measurement and signature intelligence may also be valuable, though these insights will likely only provide ancillary insights for the assessment of most military operations.

The information produced by the intelligence community can be divided into three general components. The first is the raw data collected by the intelligence community, whether it be a debrief of a human source (HUMINT), the recording of an intercepted phone call and associated metadata for that call (SIGINT), or a curated set of satellite imagery with notations from the intelligence analyst (GEOINT). The second component is the intelligence information report (IIR) produced from these data—the IIR is a widely disseminated "raw intelligence report" that provides an analyst's initial summary of the key insights from this data in a standardized, though largely unstructured, text format.[5] The third is final, evaluated intelligence that combines and cross-validates reporting from a variety of different sources to provide insights on critical information requirements from the commander.

As we summarize in Table 3.1, the SML approach described in Chapter Two can be used to derive assessment-relevant insights for a variety of common campaign objectives from the unstructured text in the IIRs produced by the intelligence community. These IIRs frequently contain specific information on the location of enemy forces; discussions of enemy capabilities, activities, plans, and morale; estimates of the ene-

---

[1] Matt Pottinger, Michael T. Flynn, and Paul D. Batchelor, "Fixing Intel: A Blueprint for Making Intelligence Relevant in Afghanistan," Center for a New American Security, 2010.

[2] Based on discussion in Appendix A of U.S. Air Force, *Annex 2-0: Global Integrated Intelligence, Surveillance & Reconnaissance Operations*, Air University, Curtis E. Lemay Center for Doctrine Development and Education, 2015.

[3] U.S. Air Force, 2015.

[4] U.S. Air Force, 2015.

[5] For example, Federal Bureau of Investigation, Directorate of Intelligence, "FBI Intelligence Information Report (IIR) Dissemination Systems (FIDS)," undated.

Table 3.1
Examples of Objective-Specific Insights Available in Intelligence Information Reports

Objective	MOE	Indicator
Degrade enemy capabilities	Enemy freedom of movement	Changes in reported location where enemy is active
	Enemy behavior	Changes in enemy plans, tactics, or morale/will to fight
	Enemy personnel and materiel	Reports of equipment and facilities destroyed, enemy attrition and desertion
Reduce population support for the enemy	Support for the enemy among the local population	Changes in reporting on local populations expressing support for the enemy
	Reduced materiel and other (e.g., intelligence, safe harbor, recruits) support to enemy from population	Enemy reports of reduction in recruitment, materiel, and other support from population
Degrade malign governance	Decreased participation and cooperation with malign governance	Enemy reports of decreased population cooperation

my's force structure; and—in some cases—enemy perspectives on the population or the populations' attitude toward the enemy. The HUMINT, SIGINT, and GEOINT data underlying these IIRs could also be analyzed using an SML approach, though this is, in practice, more complicated to implement as a result of either restrictions on data availability or the requirement for more-complex computational tools (to analyze imagery and video).

The following subsections provide three specific examples of how the SML approach could be used for the objectives and associated MOEs provided in Table 3.1. We focus on MOEs commonly used by assessment teams: (1) enemy freedom of movement, (2) size of enemy forces, and (3) enemy tactics, techniques, and procedures (TTPs). For each, we provide sample coding criteria (Step 2 of the five-step process described in Chapter Two) and describe an approach for transforming the output from the ML algorithm into an assessment-ready format (Step 5 of the five-step process).

The estimates of changes in enemy characteristics developed using this ML-based approach will be comparable, though cruder, than those developed by military and intelligence analysts.[6] The benefit of the approach described here is that it allows for objectivity and statistical relevance in examining changes over time, and for the automated processing of potentially thousands (or more) of pieces of intelligence reporting. It is not intended to replace comparable manual estimates by intelligence analysts, though—while beyond the scope of this current analysis—the insights from this auto-

[6] Estimates of each are frequently included as part of the development of an operational plan (see Appendixes A and B in Joint Publication 5-0, 2017, as an example).

mated and systematic process may provide a valuable source of information for such analysts.[7]

Further, while this approach for analyzing intelligence data can provide a range of assessment-relevant insights, the availability of data and associated IIRs is contingent on the availability of intelligence-collection assets (e.g., HUMINT collectors or technical systems) that are tasked in ways that produce assessment-relevant data. Intelligence collection is not driven by assessment needs, but rather is designed to inform ongoing operations and risks to those operations. For this reason, the volume and focus of intelligence collection on assessment-relevant characteristics may change over time, and all counts or trends obtained from these data should be scaled by volume of collection.

Example 1: Enemy Freedom of Movement

The geographic extent of the enemy's freedom of movement can be estimated using the detailed geographic information included in intelligence data. The most straightforward approach, and that recommended for assessment teams, is to compare—over time—*all* reporting that is descriptive of enemy activity or presence.

The coding criteria is relatively simple: The assessment team identifies all reporting that (1) references enemy activity or presence and (2) includes precise location information (such as military grid reference system coordinates). Transforming these data into an assessment-ready format requires only that the assessment team extract the location information from the reports identified by the ML algorithm. This locational data can then be used to estimate the geographic extent of the freedom of movement of the enemy during a given reporting period. A similar approach might be used to measure enemy territorial *control*, though IIRs—given the fact that they are "raw intelligence"—usually do not provide a reliable understanding of enemy territorial control that is separate from enemy presence.[8]

---

[7] It is possible that these machine-guided analytics could support intelligence analysis, as they allow for rapid and systematic examination of large amounts of intelligence data. However, an assessment of the value of our approach to intelligence analysts is beyond the scope of this report.

[8] A number of academic studies have attempted to estimate enemy territorial control using enemy attack data (e.g., Ran Tao, Daniel Strandow, Michael Findley, Jean-Claude Thill, and James Walsh, "A Hybrid Approach to Modeling Territorial Control in Violent Armed Conflicts," *Transactions in GIS*, Vol. 20, No. 3, 2016, pp. 413–425; Therese Anders, Hong Xu, Cheng Cheng, and T. K. Satish Kumar, "Measuring Territorial Control in Civil Wars Using Hidden Markov Models: A Data Informatics-Based Approach," 31st Conference on Neural Information Processing Systems, 2017). An IIR-based version of their approach could be implemented by parsing the coding criteria to separately identify reports mentioning enemy presence, gains, and losses. However, the accuracy of the results from this analysis would face the same challenges as those from the enemy attack data, as such analysis requires the assessment team to make relatively strong assumptions about what indicates a "gain" or a "loss." In addition, the ML-based analysis would also require that the language used to described gains and losses remains relatively similar across the data, which is unlikely to be the case as the circumstances surrounding that gain and loss are likely to be very heterogeneous.

Example 2: Size of Enemy Force that is Active

Intelligence reporting contains detailed information on the size of enemy forces. Specifically, though unlikely to contain accurate estimates of the total size of the enemy force, given the inherent difficulty in estimating the size of irregular enemy forces, intelligence reporting can provide a systematic understanding of the number of enemy fighters that are active at a given time.

The coding criteria is again straightforward, with the assessment team identifying any reports mentioning the size of enemy forces. Though substantially more difficult than other post-ML analyses discussed in this report, methods developed to estimate animal population size based on "opportunistic" data collection by civilian observers can be used with these enemy count data to estimate the overall population of the enemy force that is active at a given time.[9] The assessment relying on these data then compares the estimated total size of the active enemy force across reporting periods.

Example 3: Changes in Enemy Behavior

Changes in enemy behavior can be measured using the detailed information on enemy TTPs available in intelligence reporting.[10] The most straightforward approach, and that recommended for assessment teams, is to document all TTPs mentioned in intelligence reporting and then examine whether there are significant changes in the frequency of different TTPs over time.

The coding criteria for this MOE are more complicated than the other two, and should be developed in a two-stage process. The first stage is a consultation with intelligence analysts to identify common terminology and language that they are actively using to describe the TTPs of the enemy. The second stage involves reviewing a sample of intelligence reports selected using this language, and then reviewing these reports in detail to develop a full taxonomy of the enemy activity and behavior of interest for assessment. The post-ML processing of these data is focused on calculating the frequency of reporting on different types of enemy TTPs, which allows the assessment team to identify whether there is a shift in enemy activity or behavior.

---

9 For example, Joshua J. Horns, Frederick R. Adler, Çağan H. Şekercioğlu, "Using Opportunistic Citizen Science Data to Estimate Avian Population Trends," *Biological Conservation*, Vol. 221, May 2018.

10 The systematic analysis of intelligence data to detect changes in enemy TTPs has been used widely for counter–improvised explosive device efforts (e.g., North Atlantic Treaty Organization, *Allied Joint Doctrine for Countering—Improvised Explosive Devices*, Allied Joint Publication 3.15(A), March 2011; U.S. House of Representatives, *The Joint Improvised Explosive Device Defeat Organization: DOD's Fight Against IEDs Today and Tomorrow*, Committee on Armed Services, Subcommittee on Oversight & Investigations November 2008).

Operational Data

Deployed military units produce a substantial volume of systematic reporting, describing what units are doing, how they are doing it, and the challenges and successes that they might encounter. In most cases, this operational reporting also provides the most granular information on the activities and proficiency of partner forces.

The type of operational reporting that is most well-suited for supporting assessments are the SITREPs routinely produced by the subordinate units of an operational-level headquarters.[11] SITREPs describe—typically on a daily or weekly basis—planned and actual activities, obstacles encountered, and, in some cases, outcomes that resulted or did not result.[12] These outcomes typically include a factual statement of what happened as a consequence of the unit's efforts and how that aligns with the operation's objectives. For subordinate elements working with a partner force, SITREPs also describe the activities of their partners and the perceived efficacy of those activities.

The SITREP format is well structured for the SML approach described in Chapter Two, as they are provided in an unstructured, but consistent, text format and are available throughout the duration of the operation. The intended purpose of SITREPs makes them particularly amenable to assessment, as they provide detailed information that is deliberately aligned with the objectives of the military operation.

There are two major ways in which the information extracted from SITREPs can support assessment. The first is by providing a detailed record of operations, describing where and when different activities occurred and the effect that a unit was trying to achieve.[13] These data are essential to the assessment process, as they provide the temporal and geographic precision in measuring activities that is frequently necessary for determining causal impact. The second way SITREPs can support assessment is by providing a detailed historical record of partner force activities, behaviors, morale, and

---

[11] In addition to SITREPs, subordinate elements also sometimes produce self-assessments. These can take a variety of forms—from congressionally mandated reporting on partner force development to one-off products designed to highlight specific activities and their perceived efficacy—and a variety of formats (e.g., Excel, PowerPoint, Word). These self-assessments do not, in general, provide a reliable source of data for assessing military operations. However, they can support assessment in two other ways. First, for some units, they can provide the assessment team an understanding of medium- and long-term persistent activities and their intent. As an example, a one-off assessment by a psychological operations unit—provided as a Powerpoint "storyboard"—might describe one of its "series," including a discussion of the target audience analysis that led to that series design, a description of the series itself, and the effects that the unit was hoping to achieve. The second is by providing a comparison point for the insights of the assessment team.

[12] SITREPs are routinely produced at multiple echelons, with higher headquarters condensing multiple tactical-level SITREPs and adding additional content describing the activities of the reporting element.

[13] Some operational headquarters will maintain a summarized version of these data in a database format, typically referred to as a "mission tracker" or something similar that tracks basic information about each operation (e.g., location, unit involved, purpose), which can be used for this purpose if available.

perceived efficacy—this is often the single most comprehensive source of information on objectives related to building partner force capabilities.

The SML approach allows the assessment team to look systematically across all available SITREPs and create structured and tabulated "diaries." Diaries created in this way (though typically using non-ML methods) are a widely used type of data, known to be reliable and allow for quantitative analysis with "great detail, accuracy and authenticity."[14] However, as with any data source, care must be exercised to understand what operational reports do and do not contain, as well as to identify areas of likely bias in both collection and presentation of information.[15] Further, operational reporting requirements may change over time based on commander preferences.

In Table 3.2, we summarize some common operational objectives for which assessment-relevant insights can be gleaned from SITREPs. This table is largely focused on objectives related to building the capabilities of partner forces, as SITREPS are frequently the best and most consistent source of information on partner forces.

The following two subsections provide examples of how our SML approach can be used with SITREPs. The first describes how a detailed historical record of the activities of subordinate units can be assembled using this approach. This is particularly valuable for the assessment team if no other centralized repository details the efforts of subordinate units. However, SITREPs can still be useful even if such a repository exists, as these repositories are typically less detailed than SITREPs, frequently incomplete, and sometimes include activities that were planned but never occurred.[16] The second focuses on how the insights that SITREPs offer on partner force activities, behavior, and morale can be extracted using our SML approach.

Table 3.2
Examples of Objective-Specific Insights Available in Situation Reports

Objective	MOE	Indicator
Build and sustain the partner force	Partner force strength and sustainability	Force strength, attrition and recruitment, equipment
	Partner force operational capabilities	Operational tempo, range of missions, battlefield competence (command and control, etc.)
	Partner force morale	Indicators of will to fight
	Partner force professionalism	Reports of partner force corruption and abuse

[14] Yvette Ciere, Debbie Jaarsma, Annemieke Visser, Robbert Sanderman, Evelien Snippe, and Joke Fleer, "Studying Learning in the Healthcare Setting: The Potential of Quantitative Diary Methods," *Perspectives in Medical Education*, Vol. 4, No. 4, 2015.

[15] These include self-serving bias, which can be offset to some degree by triangulating with other data sources.

[16] Authors' personal observation.

A central challenge in using SITREPs for assessment is that the most detailed, tactical-level reporting is rarely aggregated in any single archive. This is unsurprising, as SITREPs are typically transmitted via email and frequently contain sensitive reporting about a unit's personnel (in addition to operational reporting). An assessment team co-located at the operational headquarters should be able to access the "raw" operational reporting from tactical elements if appropriately coordinated in advance. If necessary, the assessment team can use aggregated reporting from higher echelons of command, though these aggregated data will provide a significantly less detailed understanding of both the activities and progress toward building partner forces.

Example 1: Activities of Subordinate Units

SITREPs are frequently the best historical source of information on the activities of subordinate units, detailing unilateral efforts of that unit, assistance to partner forces (e.g., training, equipping), and partnered operations or other close mentoring of the partnered force. The coding criteria for identifying these activities should be developed inductively based on a review of the SITREPs, with the coding criteria developed separately for each type of subordinate unit as the types of activities they are involved in can differ dramatically. For example, the SITREP from a psychological operations unit will be different from an infantry unit in that both the types of unilateral activities and potential partnered forces differ. However, in all cases, the coding criteria should focus on reporting that includes specific geographical information, which is critical for supporting the assessment process.

The assessment-ready version of the output from this analysis is a structured database that describes the time, location, and (if applicable) partner force of each major activity conducted by a military unit. If possible, this database would also describe the level of effort exerted by the unit. This SITREP-derived "mission tracker" will provide the inputs against which all the other MOEs are assessed.

Example 2: Partner Force Activities, Behavior, and Morale

The SITREPs of units working with partner forces typically contain detailed reporting on the activities, behavior, and morale of these forces that can be extracted using an SML approach. Longitudinal analysis of these partner force characteristics can then be used to assess the effectiveness of partnering efforts.

The appropriate coding criteria and analytical approach for partner force–related objectives will depend on the specific objectives of the military operations. However, the following three examples provide sample coding criteria for different aspects of partner force development, which might be adapted to other partner force–related objectives:

1. **Freedom of movement of partner forces:** This is analogous to enemy freedom of movement (see discussion in the "Intelligence" Data section above), though

the coding criteria would now include SITREPs with discussion of partner force activities and precise locational information.

2. **Partner force capabilities:** This is comparable to measuring enemy behavior (see discussion in "Intelligence Data" section), but instead focused on the behavior of partner forces. The coding criteria should therefore be developed in two stages—beginning with a list of standard lexicon for operational activities (e.g., *patrol*, *operation*) associated with partner forces and then reviewed to develop a full taxonomy of terminology used in the SITREPs to describe partner force activities. The transformation into an assessment-ready format would then focus on capturing the evolving frequency of different types of activities.
3. **Morale of partner forces:** Measuring morale or the will to fight of partnered units is substantially more complex than measuring territorial influence or capabilities, as it requires information on the unit's leadership, support, cohesion, and operational competence.[17] An assessment team seeking to track will to fight should develop coding criteria for (1) the frequency with each of the unit-specific characteristics detailed in Connable et al.'s *Will to Fight: Analyzing, Modeling, and Simulating the Will to Fight of Military Units* and (2) the tone with which these units describe them.[18] The transformation into an assessment-ready database would then focus on tracking changes in each of these underlying signals that reflect partner force will to fight.

Ambient Data

In addition to the data collected by units involved in military operations, data produced and collected by other entities can provide valuable assessment-relevant insights. Most prominently, this includes social media (such as Twitter and YouTube), traditional media (such as newspapers), and imagery collected by overhead satellites, both commercial and governmental. These data, which we collectively refer to as "ambient data," can provide valuable insights about the local population (e.g., sentiment, economic conditions) that are generally complementary with survey-based methodologies commonly used for assessment.[19]

Both social and traditional media—given the prevalence of unstructured text in both—can be leveraged for assessment using our SML approach. Satellite imagery can

[17] See Figure 2.17 in Connable et al., 2018, p. 83.

[18] See discussion in Connable et al., 2018, pp. 63–82.

[19] Survey-based methodologies—polling, interviews, focus groups, etc.—are a common approach for evaluating population-focused conditions and are a valuable source of insight if resources and security conditions permit. However, these survey-based approaches are not without challenges, not least of these accessing and establishing rapport and reliable responses on sensitive topics from a citizenry that is often afraid of multiple actors in the battlespace.

also be a powerful tool for supporting assessment and has been used in recent years to measure economic activity,[20] industrial activity,[21] agricultural activity,[22] and movements of people.[23] However, available satellite imagery data are either already available in a structured format that can be analyzed directly by the assessment team or require more-sophisticated ML approaches than those delineated in Chapter Two.[24]

Social media are seeing increased use in assessment by both civilian and military analysts, as it allows analysts to examine the "behavior and attitudes" of targeted populations.[25] Social media platforms produce real-time "crowd-sourced" insights on emerging phenomena.[26] These data have been used for national- and regional-level political analysis, to characterize terrorist attacks,[27] and to collect data on otherwise difficult-to-reach populations.[28]

---

[20] Analysts have used electrification, which can be measured using the "nightlights" data produced by the National Oceanic and Atmospheric Administration that capture the quantity of light observable from space at night, as a proxy for economic activity (J. Vernon Henderson, Adam Storeygard, and David N. Weil, "Measuring Economic Growth from Outer Space," *American Economic Review*, Vol. 102, No. 2, 2012).

[21] Industrial activity can be estimated using the U.S. Geological Survey's Landsat satellite, which can be used to detect the heat signature of industrial activity (e.g., Eric Robinson, Daniel Egel, Patrick B. Johnston, Sean Mann, Alexander D. Rothenberg, and David Stebbins, *When the Islamic State Comes to Town: The Economic Impact of Islamic State Governance in Iraq and Syria*, Santa Monica, Calif.: RAND Corporation, RR-1970-RC, 2017, pp. 32–34).

[22] Agricultural activity can be estimated using the U.S. Geological Survey's Landsat satellite, which can detect photosynthetic activity based on light signature (e.g., Robinson et al., 2017, p. 32).

[23] Budhendra Bhaduri, Edward Bright, Phillip Coleman, and Marie L. Urban, "LandScan USA: A High-Resolution Geospatial and Temporal Modeling Approach for Population Distribution and Dynamics," *GeoJournal*, Vol. 69, No. 1, 2007, pp. 103–117.

[24] For a review of available prestructured satellite-imagery data and their use in assessment, see Robinson et al., 2017, and Daniel Egel, Charles P. Ries, Ben Connable, Todd C. Helmus, Eric Robinson, Isaac Baruffi, Melissa A. Bradley, Kurt Card, Kathleen Loa, Sean Mann, Fernando Sedano, Stephan B. Seabrook, and Robert Stewart, *Investing in the Fight: Assessing the Use of the Commander's Emergency Response Program in Afghanistan*, Santa Monica, Calif.: RAND Corporation, RR-1508-OSD, 2016.

[25] Tyler H. McCormick, Hedwig Lee, Nina Cesare, Ali Shojaie, and Emma S. Spiro, "Using Twitter for Demographic and Social Science Research: Tools for Data Collection and Processing," *Sociological Methods and Research*, Vol. 46, No. 3, August 2017, p. 1.

[26] In complex environments, the insights from these data can be used to more efficiently task other assets (such as aerial surveillance systems). For an example, see Guido Cervone, Elena Sava, Qunying Huang, Emily Schnebele, Jeff Harrison, and Nigel Waters, "Using Twitter for tasking remote-sensing data collection and damage assessment: 2013 Boulder flood case study," *International Journal of Remote Sensing*, Vol. 37, No. 1, 2016.

[27] Christopher A. Cassa, Rumi Chunara, Kenneth Mandl, and John S. Brownstein, "Twitter as a Sentinel in Emergency Situations: Lessons from the Boston Marathon Explosions," *PLOS Currents*, 2013.

[28] For example, Jes A. Koepfler and Kenneth R. Fleischmann, "Studying the Values of Hard-to-Reach Populations: Content Analysis of Tweets by the 21st Century Homeless," *Proceedings of the 2012 iConference: Association for Computing Machinery*, 2012.

Of the social media platforms available, Twitter is most often used by analysts because the majority of data are public and accessible.[29] Military analysts have determined that Twitter can be exploited to measure population sentiment in conflict zones, including population support for (or opposition toward) extremist groups and the national government.[30]

Some weaknesses and potential biases in social media data are worth noting. While social media initially promised an "uncensored" look into public citizens' sentiment and opinions, bias in social media datasets due to user preference and technology use have long plagued such analysis.[31] Additionally, such platforms are increasingly populated by automated accounts,[32] and groups are actively using these in disinformation campaigns.[33] Furthermore, users are increasingly self-censoring or restricting access to personal accounts because of privacy concerns,[34] and many platforms are restricting or shutting down access to data entirely.[35] With the proliferation of automated accounts, the authenticity and source of subjective opinions also present additional challenges.[36]

Traditional media, such as newspapers and magazines, are a second ambient data source that has been widely used by academics in assessment. Historically, these data have been primarily used to analyze sentiment toward economic and political issues.[37] However, in recent years, these data have been used extensively to study conflict, as traditional media often provide detailed information on the location and intensity of con-

---

[29] McCormick et al., 2017, pp. 15–16.

[30] Andrew K. Bourret, Joshua D. Wines, and Jason M. Mendes, "Assessing Sentiment in Conflict Zones Through Social Media," Monterey, Calif.: Naval Postgraduate School, 2016.

[31] For example, Luke Sloan, Jeffrey Morgan, Pete Burnap, and Matthew Williams, "Who Tweets? Deriving the Demographic Characteristics of Age, Occupation and Social Class from Twitter User Meta-Data," *PLOS One*, Vol. 10, No. 3, 2015.

[32] Zi Chu, Steven Gianvecchio, Haining Wang, and Sushil Jajodia, "Detecting Automation of Twitter Accounts: Are You a Human, Bot, or Cyborg?" *IEEE Transactions on Dependable and Secure Computing*, 2012.

[33] Nabih Bulos, "Libya's Two Wars: One on the Battlefield and One on Facebook," *Los Angeles Times*, June 17, 2019.

[34] Casey Fiesler, Michaelanne Dye, Jessica L. Feuston, Chaya Hiruncharoenvate, Clayton J. Hutto, Shannon Morrison, Parisa Khanipour Roshan, Umashanthi Pavalanathan, Amy S. Bruckman, and Munmun De Choudhury, "What (or who) Is Public?: Privacy Settings and Social Media Content Sharing," *Proceedings of the 2017 ACM Conference on Computer Supported Cooperative Work and Social Computing*, 2017.

[35] For example, see Josh Constine, "Facebook Restricts APIs, Axes Old Instagram Platform Amidst Scandals," *TechCrunch*, 2018.

[36] For example, see Chu et al., 2012.

[37] As examples, Hopkins, Kim, and Kim, 2017, discuss the use of newspaper reporting to measure public perceptions of the economy, and Krippendorff, 2004, provides a background discussion of the quantitative analysis of newspaper reporting (Daniel J. Hopkins, Eunji Kim, Soojong Kim, "Does Newspaper Coverage Influence or Reflect Public Perceptions of the Economy?" *Research and Politics*, 2017; Klaus Krippendorff, *Content Analysis: An Introduction to Its Methodology*, Newbury Park, Calif.: SAGE Publications, Inc., 2004).

flict events and the antagonists in those conflicts. The conflict data derived from this newspaper reporting have been used to assess the efficacy of military operations,[38] the quality of deployed forces in a variety of recent military operations,[39] and the impact of cellphone technology on conflict.[40] These conflict data are typically derived from a human-curated review of published newspaper reporting,[41] though analysts in recent years have begun employing ML algorithms to systematically review large quantities of conflict-related newspaper reporting.[42]

In Table 3.3, we summarize some common operational objectives for which assessment-relevant insights can be gleaned from social and traditional media. The comparative advantage of these data sources is the unique insights that they provide on the local population, either self-expressed beliefs (social media) or reporting on the local population's beliefs and behaviors (traditional media). In addition, both data sources can provide insights on the activities of partnered forces. The following two subsections provide examples of how our SML approach can be used to estimate the population's perceptions of enemy forces and local security conditions based on these data.

The use of social and traditional media for assessment—while appealing because of availability and the unique population-centric perspectives offered—is not without its challenges. For social media, the primary challenge is one of access,[43] which has become even more difficult in recent years as a result of legal action, public scrutiny, and privacy concerns.[44] Bias is a challenge for both social and traditional media, as social media do not provide a representative sample of a population, and traditional

---

[38] As an example, Johnston and Sarbahi, 2016, use these data analyze U.S. operations in Pakistan (Patrick B. Johnston and Anoop K. Sarbahi, "The Impact of US Drone Strikes on Terrorism in Pakistan," *International Studies Quarterly*, Vol. 60, No. 2, June 1, 2016).

[39] As an example, Haass and Ansorg, 2016, have used these data to analyze the quality of United Nations troops (Felix Haass and Nadine Ansorg, "Better Peacekeepers, Better Protection? Troop Quality of United Nations Peace Operations and Violence Against Civilians," *Journal of Peace Research*, Vol. 55, No. 6, 2018).

[40] For example, Jan H. Pierskalla and Florian M. Hollenbach, "Technology and Collective Action: The Effect of Cell Phone Coverage on Political Violence in Africa," *American Political Science Review*, Vol. 107, No. 2, 2013, pp. 207–224.

[41] Two leading examples of news-dependent conflict databases are the event-level Armed Conflict Location & Event Data Project (ACLED, "Methodology," webpage, undated) and UCDP Georeferenced Event Dataset (Mihai Croicu and Ralph Sundberg, "UCDP Georeferenced Event Dataset Codebook Version 18.1," Department of Peace and Conflict Research, Uppsala University, 2018; Ralph Sundberg and Erik Melander, 2013, "Introducing the UCDP Georeferenced Event Dataset," *Journal of Peace Research*, Vol. 50, No. 4, pp. 523–532).

[42] Hannes Mueller and Christopher Rauh, "Reading Between the Lines: Prediction of Political Violence Using Newspaper Text," *Empirical Studies of Conflict*, Working Paper No. 2, 2016.

[43] Some services (e.g., Twitter) provide limited public-access downloads of data, but systematic extraction of all data relevant to an ongoing military operation is becoming difficult.

[44] For example, see Amy Collins, "Your Questions Answered: Instagram's API Changes," *Brandwatch*, April 3, 2018.

Table 3.3
Examples of Objective-Specific Insights Available in Ambient Data

Objective	MOE	Indicator
Build and sustain the partner force	Partner force professionalism	Reports on the conduct of partner forces
Degrade enemy capabilities	Enemy freedom of movement	Discussions of liberation and movement of forces
	Enemy capabilities	Discussions of the efficacy and threat
	Enemy personnel and materiel losses	Reports of loss of morale, battlefield damage
Reduce population support for the enemy	Support for the enemy among the local population	Changes in reporting on local populations expressing support for the enemy
Build effective local governance	Increased population expressions of local government support	Positive reports or posts about local government and government services
Degrade malign governance	Reduced population expressions of malign governance support	Negative reports or posts concerning malign governance actions

media are necessarily selective in what they report and frequently have incomplete information about unfolding events.[45] Another challenge facing both social and traditional media is a lack of temporal and geographic specificity, as the location information available in social media data is often inaccurate and traditional media often recycle historical material.[46] For satellite-based imagery—the class of imagery data most amenable to supporting assessments using current technologies—the central challenge is that such data are typically less informative in areas that are either very densely or very sparsely populated.[47]

Example 1: Local Perceptions of the Enemy

Both social and traditional media reporting can contain information on local perceptions and support for the enemy, which typically includes a blend of positive and nega-

---

[45] For a review of this bias in traditional media for conflict-event data, see Nils B. Weidmann, "A Closer Look at Reporting Bias in Conflict Event Data," *American Journal of Political Science*, Vol. 60, No. 1, January 2016. For social media, this lack of representation limits the ability to make defensible inferences (Maeve Duggan and Joanna Brenner, "The Demographics of Social Media Users—2012," *Pew Research Center's Internet and American Life Project*, 2013), and the detailed demographic necessary to account for this reporting bias are frequently missing or unreliable (Sloan et al., 2015).

[46] For a discussion of the "spoofing" of location information in social media, see Bo Zhao and Daniel Z. Sui, "True Lies in Geospatial Big Data: Detecting Location Spoofing in Social Media," *Annals of GIS*, Vol. 23, No. 1, 2017. And for a discussion of recycling of traditional media reporting, see Benjamin D. Horne and Sibel Adali, "An Exploration of Verbatim Content Republishing by News Producers," *arXiv*, 2018.

[47] For a discussion of these challenges, see Charlotta Mellander, José Lobo, Kevin Stolarick, and Zara Matheson, "Night-Time Light Data: A Good Proxy Measure for Economic Activity?" *PLOS One*, 2015.

tive discussions—particularly if the enemy, host-nation forces, and others are actively using these platforms for information operations. The most straightforward approach, and that recommended for assessment teams, is to extract all enemy-focused reporting; develop coding criteria so that the ML algorithm can be used to categorize each report as "positive," "negative," or "neutral"; and then compare the frequency of reporting of each of these categories over time. An alternative to this second step would be to use "semantic analysis," an analogous unsupervised ML approach.[48]

Example 2: Perceptions of Security Conditions

Ambient data can similarly provide a valuable perspective on local security conditions and other types of local environmental conditions (e.g., economic opportunities, government effectiveness). Though relatively straightforward to assess using these data, a two-step SML process is required, with the assessment team developing training data for each step. The coding criteria for the first step in this process are simple, in that the assessment team need only identify reporting that discusses the type of local environmental condition relevant to the assessment (e.g., security). In the second step, the assessment team develops coding criteria to determine whether the reporting is "positive," "negative," or "neutral"—this is analogous to the semantic analysis described in the previous example ("Local Perceptions of the Enemy").

Data Discovery and Preparation

This section describes processes that the assessment team can follow for discovering and preparing the intelligence, operational, and ambient data described in this chapter. While these data sources will be familiar to most analysts, the extraction of large quantities of data—rather than, for example, a single intelligence report—to a format that is amenable to SML is likely a bit less familiar.

Figure 3.1 provides an overview of the major steps in this process for each of the data sources. Though each process is different—a consequence of the structural differences in these data and the systems that archive them—the goal in each case is to extract all potentially relevant data for the area in which the operation is occurring.

Intelligence

There are a wide variety of systems designed to facilitate discovery of intelligence (e.g., Multimedia Message Manager, Library of National Intelligence), each of which typically allow analysts to use location, time, and Boolean logic to discover relevant report-

48 For a canonical example, see Peter D. Turney, "Thumbs Up or Thumbs Down? Semantic Orientation Applied to Unsupervised Classification of Reviews," *Proceedings of the Association for Computational Linguistics*, 2002. This approach is unsupervised in that it does not require the assessment team to code reporting—the downside is that it requires training on a second and slightly more complicated analytical tool.

Figure 3.1
Data Preparation Workflow

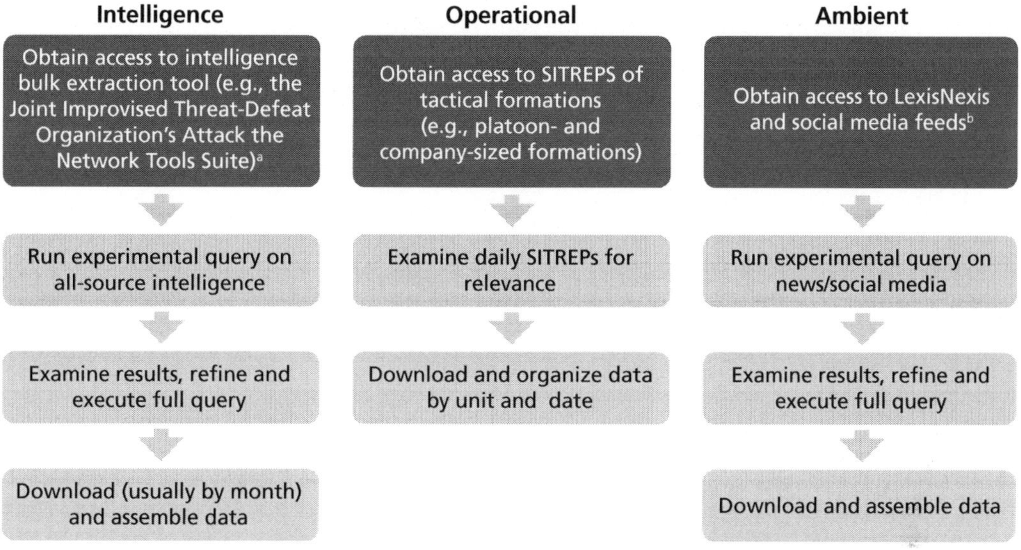

[a] Name of intelligence platform current as of winter 2019. Other all-source intelligence-focused platforms with unrestricted download capacity may be available in the future.
[b] LexisNexis provided the most intuitive platforms for traditional media at the time of this writing in mid-2019. New platforms may be available.

ing. However, few of these tools allow analysts to export the intelligence reporting in a bulk, organized format amenable for SML. Indeed, most tools do not allow exportation of large quantities of reporting—thousands or tens of thousands of reports for a given operational area is not uncommon—and fewer yet allow export of the full text of all those reports.

One system that does allow this necessary bulk extraction of intelligence reporting is the Joint Improvised Threat-Defeat Organization's Attack the Network Tools Suite (ANTS).[49] This system allows analysts to identify all available reporting from "more than 550 sources across all the combatant commands and the intelligence community" and then extract the full text of all reports in a database format amenable to analysis by the SML methods.[50] Data discovery is possible via a number of ANTS tools, and ANTS's Catapult tool provides the download capability.

Data discovery via the Catapult tool is straightforward, as it allows for searching based on Boolean terms, geography, and time of reporting. We recommend that the assessment team begin by selecting all relevant countries for the operations, the full

---

[49] The Combined Information Data Network Exchange (CIDNE) provides an analogous capability, though is not available for all geographic areas.

[50] Patrick Tucker, "The Apps They Carried: Software, Big Data, and the Fight for Mosul," *DefenseOne*, October 17, 2016.

time frame for the operation, and the name of the enemy force—and any potential acronyms—as the Boolean search terms. This will include significant numbers of irrelevant reporting, because the geographic search parameter includes all reporting that even just mentions a country. This irrelevant reporting will be removed as part of the SML analysis, though refining the search parameters to be as careful as possible will improve the performance of the SML algorithms.

Data preparation is also straightforward, as the Catapult tool has an export feature that allows the analyst to export all data as a structured database. However, the data files produced using this feature can be very large given the quantity of data being extracted. Thus, in practice, it is recommended that the assessment team download a single month of data at a time, or perhaps a week, depending on the volume of available intelligence and the capabilities of available computer systems.

Operational Data

SITREPs are regularly produced, typically either daily or weekly, by all echelons of command. Ideally, the assessment team would have access to SITREPs produced by every command echelon, though the detailed and granular reporting in the SITREPs of tactical elements is likely the most useful for the assessment team.

Accessing all relevant SITREPs for a given operation is straightforward in principle, but more difficult to do in practice. In particular, while SITREPs for the operational-level headquarters will typically be readily available to the assessment team—often via a shared drive or portal—the SITREPs of subordinate elements are less accessible, particularly those two or more echelons below the headquarters. Coordinating access to SITREPs of subordinate elements is possible, as these are typically disseminated via email or a similar format, though this will require coordination between the assessment team and the headquarters' operations directorate (J-3).

Preparing the SITREP data for analysis using SML is similarly straightforward, as the assessment team need only convert the operational reporting into an Excel file with the unit's name and date. This might be tedious, and the process could potentially be automated, but reporting can change somewhat over time, and a manual review of the SITREPs while organizing the data can help identify emerging issues worthy of inclusion in the assessment.

Ambient Data

There are a variety of existing tools that can be used to discover and prepare traditional and social media for analysis, as these data are widely used by both academic and private-sector researchers. For traditional media, one widely used tool is Lexis-Nexis, which offers a platform that is intuitive, provides access to current and historical reporting in thousands of global newspapers, and allows analysts to download relevant reporting in an easily analyzed database format. For social media, the only widely available data come from Twitter, which provides researchers the ability to purchase both current and historical data.[51]

Geography-based discovery is possible, though more difficult, for ambient data than for either intelligence and operational data. For traditional media, the easiest way to do geography-based discovery is to limit the sources being consulted—for example, constraining the search application to include only publications from within the countries affected by the operation.[52] However, for social media, this is generally not recommended, because geographic information is typically available for only a very small share of available reporting, and these are unlikely to be representative.[53]

Text-based discovery using Boolean logic is thus critical for both traditional and social media. Selecting ideal search parameters will typically require some experimentation, based on the density of available reporting. In general, we recommend starting with a wide aperture—including all newspaper articles referencing the country or any of the antagonists in the conflict (both U.S. partners and otherwise). For denied environments (e.g., operations in rural parts of the Democratic Republic of the Congo), this may produce a manageable number of reports to analyze via the SML approach, as SML provides a more intelligent way of filtering reports than Boolean logic. However, the assessment team should experiment to determine whether it is possible to refine the Boolean search terms without eliminating potentially useful reporting, as a more targeted search will improve the efficacy of SML.

Data preparation is straightforward, as the available commercial tools typically allow export in a database format that can be analyzed directly. However, because duplication of reporting is very common in both traditional and social media—for example, newspapers often copy text verbatim, Tweets are often reposted—it is important to select the appropriate download "options" that allowing tracking of duplicates.

Integrating Data into Operation Assessment

The structured data produced using these SML approaches are not the end product in itself. Rather, these structured data must be integrated into an overall assessment process. Though designed to support the standardized assessment process for SOF

---

[51] See discussion at Adam Tornes, "Introducing the Premium Full-Archive Search Endpoint," Twitter Developer Blog, February 1, 2018.

[52] An alternative is to "georeference" the articles based on the locations mentioned in the text, which is now possible with reasonable accuracy though requires specialized tools (Hao Chen, Maria Vasardani, and Stephan Winter, "Georeferencing Places from Collective Human Descriptions Using Place Graphs," *Journal of Spatial Information Science*, Vol. 17, 2018).

[53] For example, less than 1 percent of Twitter data include geocoordinates (David Jurgens, "That's What Friends Are for: Inferring Location in Online Social Media Platforms Based on Social Relationships," *Proceedings of the 7th International Conference on Weblogs and Social Media*, 2013).

that was the subject of our previous report,[54] these structured data are appropriate for a variety of operation assessment approaches. This structure allows analysts to link these data to the activities of the operational-level headquarters and to triangulate the findings with other types of structured data. In the following chapter, we illustrate this process.

---

[54] Robinson, Egel, and Brown, 2019.

CHAPTER FOUR

Example Using Operation Observant Compass

This chapter illustrates the integration of our SML approach into the assessment process, using the recently concluded campaign against the Lord's Resistance Army (LRA)—Operation Observant Compass—as an example. Our intent is to demonstrate how the insights extracted from intelligence, operational, and ambient data using SML tools can provide findings that are objective and statistically relevant. This demonstration both describes the creation of assessment-relevant databases, following the approach described in Chapters Two and Three, and illustrates how existing analytical tools can be used to incorporate these data into assessment.

We selected Operation Observant Compass as an example for three primary reasons. The first is that the campaign included enemy-centric, partner-centric, and population-centric operational objectives, making each of the three classes of data discussed throughout this report relevant. The second is that it provides an example of SML in a relatively "data-poor" environment; the operational-level headquarters was not adequately resourced to collect assessment-specific data (e.g., polling), and the volume of ambient data is limited. Operation Observant Compass therefore provides an example of how SML-derived insights can satisfy data requirements of the assessment team and demonstrates that these tools can provide valuable assessment-relevant insights even when data are limited.[1] The third reason we selected Operation Observant Compass as an example is that it was a long-duration mission, demonstrating the value of these tools in providing an objective measure of long-term changes in conditions.

Our discussion in this chapter is organized by the three classes of data discussed throughout this report rather than by the operational objectives of Operation Observant Compass, as would normally be done for a campaign assessment. The exemplar MOEs that we provide for each data type, though relevant to Operation Observant Compass, were selected for illustrative purposes only and are not comprehensive. That said, each class of data roughly corresponds to a single class of operational objectives

[1] The tremendous volume of reporting in "data-rich" environments (e.g., operations in developed country contexts) can create its own challenges—particularly if adversaries are using these platforms for disinformation—though the procedures described in this report are still broadly applicable.

(e.g., intelligence data are most applicable to enemy-centric objectives), and the MOEs presented in this discussion were selected based on their applicability to this campaign and their broad applicability to a wide variety of other campaigns.

This chapter divided into five sections. The first describes Operation Observant Compass and its key lines of effort. The next three sections detail the use of SML tools to analyze the three classes of data discussed throughout this report: intelligence, operational, and ambient. The intelligence and operational data discussions in this unclassified report use an unclassified (ambient) data source as proxy for the actual classified intelligence and operational data; this allows us to fully describe our approach in this report. A classified appendix provides the full discussion using the actual classified reporting, and Appendix B of this report provides the computer code that was used to produce these findings.

We then describe how analogous methods can be used to develop a systematic database of U.S. activities, which is critical for assessment, when those data are not otherwise available. Finally, we discuss how these insights can be combined into a "Commander's Assessment Dashboard" that can provide the commander up-to-date insights on the effectiveness of forces.

Background on Operation Observant Compass

In May 2010, the United States committed to "support multilateral efforts to successfully protect civilians and eliminate the threat posed by the Lord's Resistance Army,"[2] a U.S.-recognized terrorist organization operating in Uganda, the Central African Republic, the Democratic Republic of the Congo, and South Sudan.[3] The following year, in October 2011, President Barack Obama announced the deployment of U.S. SOF to "act as advisors to partner forces that have the goal of removing from the battlefield Joseph Kony and other senior leadership of the LRA."[4]

---

[2] Specifically, the United States committed to "providing political, economic, military, and intelligence support for viable multilateral efforts to protect civilians from the Lord's Resistance Army, to apprehend or remove Joseph Kony and his top commanders from the battlefield in the continued absence of a negotiated solution, and to disarm and demobilize the remaining Lord's Resistance Army fighters," in addition to targeted humanitarian and reconstruction assistance (Public Law 117-172, Lord's Resistance Army Disarmament and Northern Uganda Recovery Act of 2009, May 24, 2010). The United States had provided support to counter-LRA efforts as early as 2008 (James J. F. Forest, *U.S. Military Deployments to Africa: Lessons from the Hunt for Joseph Kony and the Lord's Resistance Army*, MacDill Air Force Base, Fla.: Joint Special Operations University, Report 14-4, 2014, p. 19).

[3] Public Law 117-172, 2010.

[4] The White House, "Letter from the President to the Speaker of the House of Representatives and the President Pro Tempore of the Senate Regarding the Lord's Resistance Army," October 14, 2011. The 100 U.S. SOF personnel that were initially committed grew to approximately 150 by 2014 (Forest, 2014, pp. 4, 73).

These U.S. SOF personnel would provide "information, advice, and assistance" to partners across the four countries,[5] though they were not authorized to "engage LRA forces unless necessary for self defense."[6] Though the U.S. effort was initially focused on killing or capturing Kony, efforts to promote LRA defections—via psychological operations (e.g., leaflets, radio), conducted both unilaterally and in partnership with partners—became a central line of effort by 2013.[7]

Counter-LRA efforts ended in April 2017, with the LRA "dramatically weakened . . . in numbers and overall effectiveness."[8] However, while these efforts had "captured four of the five key LRA leaders" and dramatically reduced the size of the LRA from 2,000 to 100 "active" members, the primary goal of the United States and its partners—the capture or elimination of Joseph Kony—had not been achieved.[9]

Intelligence Data

Given the focus on degrading LRA senior leadership, the intelligence data available for Operation Observant Compass provide relatively detailed information on the LRA. This includes both reported locations of LRA forces—typically referenced in the context of a suspected attack, movement, bed-down location, or facilities—and their evolving TTPs. These data can be used as indicators to measure changes in enemy freedom of movement and enemy behavior—MOEs that are relevant to Operation Observant Compass and other campaigns in which degrading enemy capability is a primary operational objective.

In this section, we describe a standardized workflow that can be used for extracting information for each of these two enemy-focused MOEs. Though the relevant intelligence data are classified, the enemy-focused and unclassified Armed Conflict Location & Event Data Project (ACLED) enable us to illustrate this workflow for the enemy freedom of movement MOE.[10] The classified appendix provides comparable analysis for both MOEs using available intelligence data.

---

[5] These four partner forces were the Ugandan People's Defense Forces (UPDF), the Sudanese People's Liberation Army (SPLA), Forces Armées Centralafricaine (FACA), and Forces Armées de la République Démocratique du Congo (FARDC).

[6] The White House, 2011.

[7] Jonathan R. Easter and Benon M. Hatangimana, *Bending the Spear: The Campaign Against the Lord's Resistance Army*, Monterey, Calif.: Naval Postgraduate School, December 2017, p. 41.

[8] U.S. Africa Command, *U.S. Forces Transition Counter-LRA Mission to Broader Security and Stability Activities*, March 29, 2017.

[9] Zack Baddorf, "Uganda Ends Its Hunt for Joseph Kony Empty-Handed," *New York Times*, April 20, 2017.

[10] ACLED, which is unclassified and publicly available, provides the location and time of violent attacks between armed groups, attacks against civilians, and a variety of other conflict-related events based, primarily, on local and international news reporting. For an overview of these data, see Clionadh Raleigh and Caitriona Dowd,

MOE 1: Enemy Freedom of Movement

Analyzing changes in the geographic dispersion of reported LRA activity provides a tool for assessing changes in the enemy's freedom of movement. We illustrate the types of insights that emerge from this analysis of enemy freedom of movement in Figure 4.1, which reports our estimates of LRA freedom of movement for three periods: (1) 2009–2011, which roughly corresponds to the period before the beginning of Observant Compass; (2) 2012–2016, the years in which Observant Compass was in full operation; and (3) 2017–2018, which roughly corresponds to the end of operations and the period immediately after. Analysis of these data for an assessment is likely to be done with some type of regular periodicity (e.g., annual), as discussed earlier in this report and later in this chapter, but these ranges were picked for illustrative purposes.

Although the structured and unclassified ACLED data used in Figure 4.1 are illustrative of the likely impact of Operation Observant Compass on LRA freedom of movement, the number of data points is limited—for example, only 319 reports are available for the five years when Operation Observant Compass was most active (2012–2016). Comparable freedom of movement can be generated from intelligence reporting using the following workflow (the classified appendix provides additional details):

1. **Curate data:** Extract all sentences that mention the LRA and have precise geographical information. Villages, roads, and other geographic features referenced in intelligence reporting are typically annotated in-line with the corresponding military grid reference system geocoordinates.
2. **Develop coding criteria:** Review 100 sentences containing both the LRA and precise locational information and develop coding criteria appropriate for freedom of movement (e.g., mentions LRA activities or facilities).
3. **Construct training data:** Apply the coding criteria to 300 sentences.
4. **Implement and calibrate:** Execute ML algorithms using the training data to identify all reporting mentioning LRA activity or facilities. Review results from ML algorithms and add additional training data or modify algorithm parameters as needed.
5. **Transform output:** Extract locational information for each identified report and build a geospatial database with the date and location of each report. This can then be used with geospatial tools to directly create the maps in Figure 4.1.

MOE 2: Enemy Behavior

Enemy TTPs are frequently discussed in intelligence reporting. This includes discussion of specific enemy TTPs currently being employed and whether there were notice-

"Armed Conflict Location and Event Data Project (ACLED) Codebook," Armed Conflict Location and Event Data Project, 2015.

Figure 4.1
Lord's Resistance Army Freedom of Movement

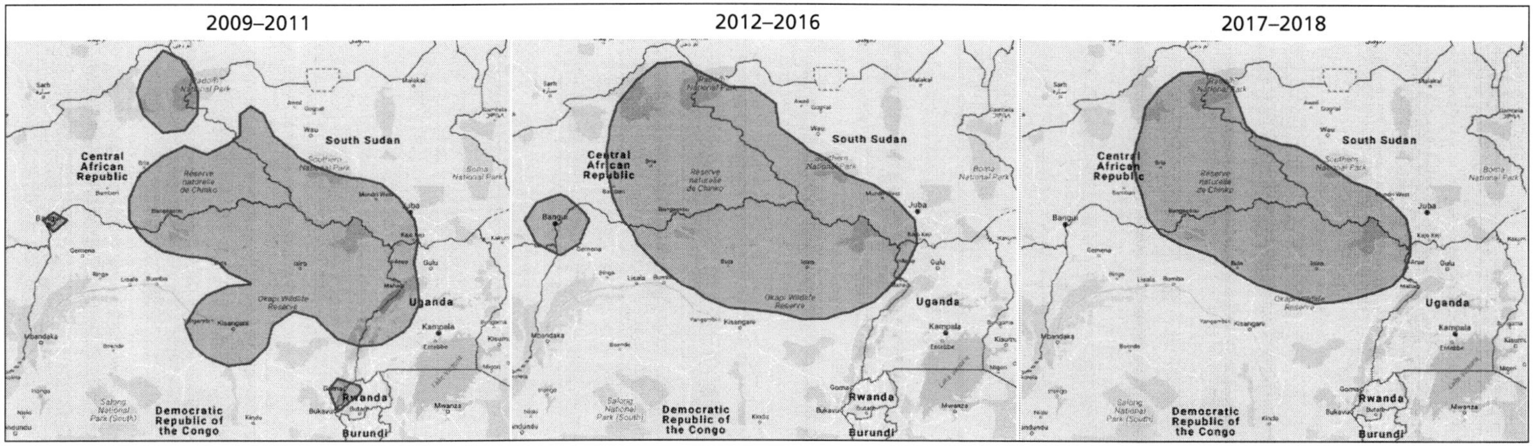

NOTES: Estimates of freedom of movement (highlighted areas) includes all ACLED reporting from the corresponding year that mentions the term LRA in its description. We leverage an existing algorithm developed by ecologists to measure the range of animals, the "95% kernel home range," to calculate the likely control of LRA for different time ranges (Mark A. Burgman and Julian C. Fox, "Bias in Species Range Estimates from Minimum Convex Polygons: Implications for Conservation and Options for Improved Planning," *Animal Conservation*, Vol. 6, No. 1, February 2003).

able changes in those TTPs. These data enable us to develop a longitudinal analysis of the prevalence of different types of LRA TTPs, and the types and timing of reported changes in TTPs. These data can be extracted as follows:

1. **Curate data:** Identify all sentences that mention "LRA," "TTP," "tactics," "techniques," or "procedures."
2. **Develop coding criteria:** Develop an LRA-appropriate taxonomy of TTPs by reviewing approximately 10 percent of reports.
3. **Construct training data:** Review reports extracted in Step 1 and code for whether a specific TTP category is included—at least 100 unique reports should be reviewed per category included in the taxonomy.
4. **Implement and calibrate:** Apply ML algorithms to identify all reports referencing each LRA TTP—individual reports may correspond to multiple TTP categories. Review results from ML algorithms and add additional training data or modify algorithm parameters as needed.
5. **Transform output:** Create counts of the number of reports mentioning each TTP per year (or per month), which can then be used for analysis in Excel or an analogous software for generating charts.

Operational Data

Operation Observant Compass was conducted largely by, with, and through local partners.[11] Thus, operational data provide extensive details on the strength and operational capabilities of partner forces as they carried out operations against the LRA, allowing assessment of the evolving capabilities of the partner forces.

In this section, we provide a workflow for how our SML approach can be used to create MOEs for partner force operational capabilities. The analysis in this unclassified report relies on newspaper reporting, which contains a sufficient amount of systematic reporting on these partnered forces for illustrative purposes, and the same approach could be applied to SITREPs produced by military units.

MOE 3: Partner Force Operational Capabilities

Reports regarding the success of partner force operations—whether through enemies killed or captured, protecting the population, or securing territory from the LRA—provide a measure of the success of U.S. forces in building and sustaining a successful partner force. We illustrate the types of insights that can be extracted from operational reporting in Figure 4.2. In this case, we have used our SML approach to estimate the number of successful U.S.–partner force engagements against the LRA,

[11] Easter and Hatangimana, 2017, p. 39.

Figure 4.2
Partner Force Success over Time

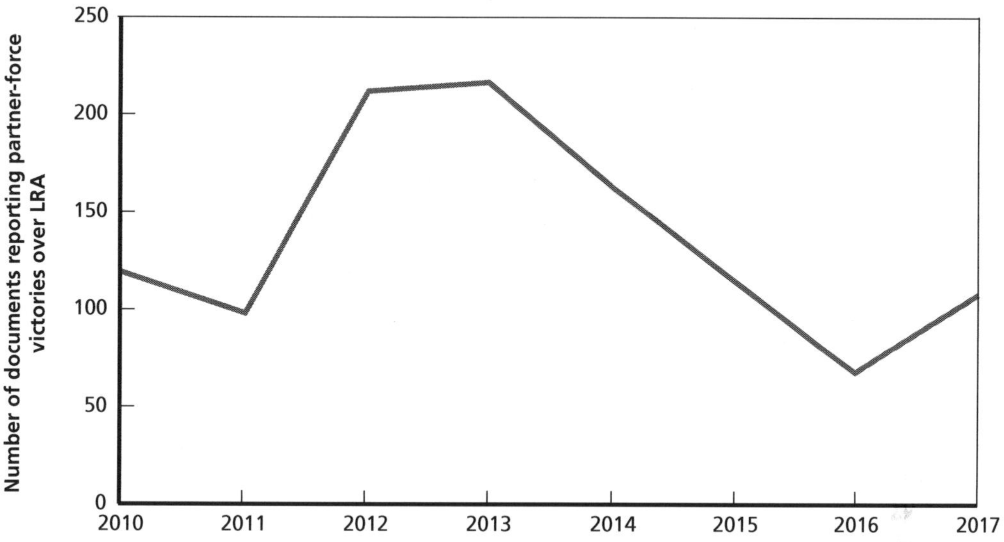

and a comparable approach could be used to estimate the number of unsuccessful engagements, the frequency of coordination among U.S. partner forces, and other partner force–focused objectives.

The following workflow describes the analytic process for extracting and analyzing reports of partner force operational successes over time. Note that media reporting is typically insufficient for providing a complete picture of partner force operational capabilities, but it can provide a useful perspective on how the host government and local population view these capabilities.

1. **Curate data:** Identify all reporting that mentions any partner force by name or acronym. In this case, partner forces included the Ugandan People's Defense Forces, or "UPDF," the Sudanese People's Liberation Army, or "SPLA," Forces Armées Centralafricaine, or "FACA," and Forces Armées de la République Démocratique du Congo, or "FARDC."
2. **Develop coding criteria:** Specify MOEs appropriate for each partner force (e.g., operational success) and review 100 reports to develop coding criteria for each MOE.
3. **Construct training data:** Review 300 sentences from curated reporting that mention at least one partner force, and construct the initial training data.
4. **Implement and calibrate:** Apply ML algorithms to identify all reports discussing operational success (or other identified MOEs).
5. **Transform output:** Restructure as longitudinal database that includes the total number of references to operational success per year (or per month).

Ambient Data

In news and other media, there was significant local and international reporting about the activities of the LRA during Operation Observant Compass. This was in large part because of the LRA's violence, which included rape, enslavement, and massacre of civilians. In addition, the Invisible Children campaign ensured a high level of public attention to counter-LRA efforts, with much of it focusing on progress toward degrading LRA leadership. Ambient data—from both traditional and social media—therefore contain a substantial amount of reporting about the perceived strength and capabilities of the LRA and about public opinions of progress toward counter-LRA goals. These data can be used as indicators to measure perceptions of the threat to civilians posed by the LRA (an "enemy capabilities" MOE) and public perceptions of progress toward degrading LRA leadership (an "enemy personnel and materiel losses" MOE).

In this section, we describe a standardized workflow for these two population-focused MOEs. The first example, focused on measuring perceptions of enemy capabilities for posing a threat, relies on news and opinion articles downloaded from LexisNexis and spanning from 2010 through 2017. The second example, focused on public perceptions of progress toward degrading LRA leadership, relies on approximately 60,000 LRA-relevant tweets posted from December 2016 through March 2017.

MOE 4: Enemy Capabilities

Reports in traditional media regarding the level of threat posed by the LRA—whether in terms of the size and capabilities of its forces, its ability to sustain and reconstitute itself, or its abilities to inflict damage—provide a measure of enemy capabilities over time.[12] We illustrate the types of insights that can be extracted from ambient data regarding enemy capabilities in Figure 4.3. In this case, we used our ML approach to identify reporting that both describes (1) reports of a strong, resilient, persistent, or growing LRA threat and (2) reports of a decreased or weakened LRA.

The following provides a workflow for extracting perceptions of the threat posed by the LRA—and specifically whether that threat is increasing or decreasing—using the newspaper reporting available in LexisNexis. Note that media reporting alone is insufficient as a measure of enemy capabilities, but it can be used to triangulate with reporting from intelligence and other data.

1. **Curate data:** Extract all reporting mentioning LRA.
2. **Develop coding criteria:** Review reporting to understand how newspapers describe the LRA threat, which includes all language that describes the tough-

[12] Civilian impressions of local security and threats—in this case the ongoing level of threat posed by the LRA—are an important source of information that can be triangulated against these more formal estimates. In some operations, increasing civilian perceptions of local security may even be a formal operational goal. For example, see Brian Petit, "The Fight for the Village: Southern Afghanistan," *Military Review*, May–June 2011.

Figure 4.3
Lord's Resistance Army Perceived Threat over Time

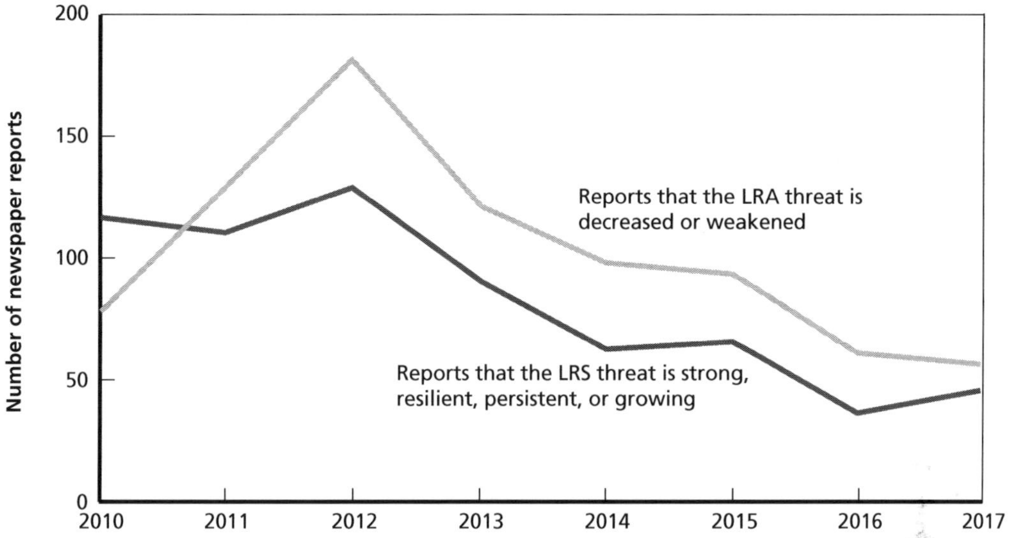

ness, resilience, strength, and operational capabilities of the LRA. In our case, we included all language referring to increased (or decreased) operational tempo, growing (or diminishing) force strength, and level of threat posed by the LRA to civilians and to regional stability.

3. **Construct training data:** Build a training dataset with two columns of information—one for whether the report mentions an increase in LRA capabilities and the other a decrease in capabilities.
4. **Implement and calibrate:** Apply ML algorithms to identify all reports referencing LRA threat increases and decreases.
5. **Transform output:** Restructure as a longitudinal database that includes the total number of references to strengthening or weakening of the LRA per year (or per month).

MOE 5: Enemy Personnel and Materiel Losses

Reports in social media regarding the degree to which the public believes counter-LRA efforts are making progress toward capturing or killing Kony and other LRA leaders provides one measure of enemy losses. In this case, we used our approach to estimate the degree to which the public (reflected by individual tweets) was optimistic or pessimistic regarding the chances of seriously degrading LRA leadership. A comparable approach could be used to assess public fear of the LRA, public impressions of civilian suffering under the LRA, or other enemy or population-focused objectives.

In Figure 4.4, we illustrate the types of insights that can be gleaned from applying our ML approach to social media—in this case, all tweets mentioning both "Kony" and the "LRA" from December 2016 to March 2017.[13] Out of the 1,069 tweets that were determined to reflect opinions on progress toward killing or capturing Kony and other LRA members, 865 tweets were either factual (e.g., retweeting articles from other sources without further comment) or had unclear or ambivalent sentiment regarding the C-LRA campaign. A further 106 tweets expressed optimism about the campaign, and 98 expressed pessimism. Thus, in this case, overall popular sentiment seemed roughly evenly split.

The following steps describe a workflow for extracting optimistic versus pessimistic opinions on progress toward degrading LRA leadership in social media. Note that Twitter reporting alone is insufficient as a measure of enemy capabilities, but it can be used to triangulate with other data.

1. **Curate data:** Extract all tweets mentioning Kony and LRA.
2. **Develop coding criteria:** Review tweets to understand how Twitter users report on counter-LRA progress. In our case, we noticed that most tweets that were recognizably optimistic or pessimistic focused on capturing or killing members of LRA leadership. Additionally, most tweets were informational in nature, retweeting articles or other content that did not take a clear subjective side.

Figure 4.4
Public Opinion on Progress Toward Degrading the Lord's Resistance Army

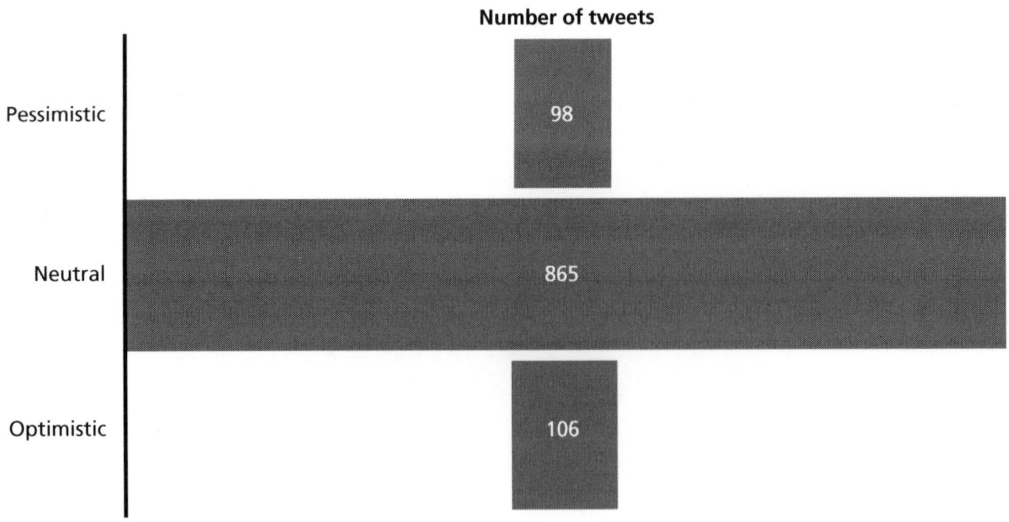

[13] Twitter data can be costly when extracted from private data providers. We analyzed a small slice of Twitter data for this reason. More-expansive access to social media would allow longitudinal analysis.

3. **Construct training data:** Build training data with two columns of information—one for whether the tweet is clearly optimistic and one for whether it is clearly pessimistic.
4. **Implement and calibrate:** Apply ML algorithms to identify all tweets referencing optimism and pessimism toward degrading LRA leadership.[14]
5. **Transform output:** Restructure as a database that includes the total number of optimistic, pessimistic, and neutral tweets.

Operational Activities

An objective and systematic measure of the activities of an operational-level headquarters is critical to measuring the effectiveness of those activities. While some headquarters maintain detailed databases tracking these activities—the "five W's" (who, what, where, when, and why) that are typically required of tactical-level mission planning[15]—this information is often available in only the SITREPs of tactical-level units. In these cases, the SML approach can be used to build a detailed activity database from these SITREPs that describes the types, timing, and intensity of activities conducted by subordinate units and partner forces.

Available SITREPs for Operation Observant Compass describe the blend of partnering and psychological operations that were critical to U.S. efforts.[16] The following workflow, which is illustrated in the classified appendix, can be used to build a database of each class of operational activity using these SITREPs:

1. **Create database of all tactical-level SITREPs from Operation Observant Compass.** This includes all subordinates units, including civil affairs, psychological operations, and SOF units.
2. **Develop coding criteria for operational activities.** This is best done iteratively, starting with general keywords that are likely to be indicative of discussions of either partnering activities (e.g., "partnering," name of partner forces, etc.) or psychological operations activities (e.g., "leaflets," "radio," "messaging,"

[14] In this case, we tested the ability of an ML algorithm to attempt to replicate a human coder's decisions using fivefold cross-validation. Because the small number of relevant observations, recall was low (19 percent and 21 percent for optimism and pessimism, respectively), but accuracy (93.5 percent and 92 percent, respectively) and precision (95 percent and 96 percent, respectively) were high. With a larger training dataset, we would expect recall to increase. See Pedro M. Domingos, "A Few Useful Things to Know About Machine Learning," *Communications of the ACM*, Vol. 55, No. 10, 2012, pp. 78–87.

[15] For example Army Techniques Publication 3-55.4, *Techniques for Information Collection During Operations Among Populations*, Washington, D.C.: Headquarters, Department of the Army, April 2016.

[16] Psychological operations have often been praised in the public media for helping to bring about the demise of the LRA through attrition of its ranks (Jared Keller, "How the Army Decimated Joseph Kony's Messianic Death Cult Without Firing a Shot," *Task and Purpose*, October 17, 2017).

"series," etc.). Consider limiting to only sentences that have precise geographical information (e.g., MGRS, city or other location identifier). This should be updated quarterly.

3. **Build training data**. Review 300 identified sentences and identify which mention each type of identified activity.
4. **Implement and calibrate:** Apply ML algorithms to extract all activities and associated information (e.g., location, time). Review results from ML algorithms and add additional training data or modify algorithm parameters as needed.
5. **Transform output:** Restructure as a database that includes the time, location, and type of activity.

Commander Decisionmaking

A core value of the SML tools is that they can be automated, thus providing near-real-time insights when integrated into a full assessment process, as described in previous RAND research.[17] Integration of SML could allow the commander to receive daily updates—comparable to the ubiquitous intelligence and operational updates—on progress toward operational objectives. We anticipate that this would stimulate a command-wide discussion that could improve the efficacy of planning and operations,[18] though we are not aware of any assessments that have deployed such an approach.

Figure 4.5 provides a sample commander's dashboard, which is a compilation of the insights from the intelligence, operational, and ambient data. In addition, these data can and should be augmented with insights emerging from available structured data, including both assessment-specific insights (e.g., from polling) or those from ambient reporting (e.g., nightlights). In this example, the commander can flexibly query the system for different visualizations of the data. In all cases, it is critical that a long-term perspective is provided so that the commander can examine long-term trends. The specifics of what types of visualization to include should be a collaboration of the command ground and the assessment team, as they will need to coordinate to ingest these data into these situational awareness tools.

In addition to these simple summary statistics, the assessment team would also include figures that provide descriptive insights into possible causal relationships between the efforts of the command and the observed change in MOEs. For example, if U.S. and partner force operational timing and locations were available via a preexist-

---

[17] Robinson, Egel, and Brown, 2019.

[18] Schroden et al., 2013, observed that designing assessments "to stimulate discussions leading directly to decisions by senior leaders on actions they could take, direct, or request" was a best practice. Though their approach for stimulating this discussion was different, we anticipate that integrating assessment into the daily battle cycle could generate discussion similar to that currently had for intelligence and operational updates.

Figure 4.5
Commander's Dashboard

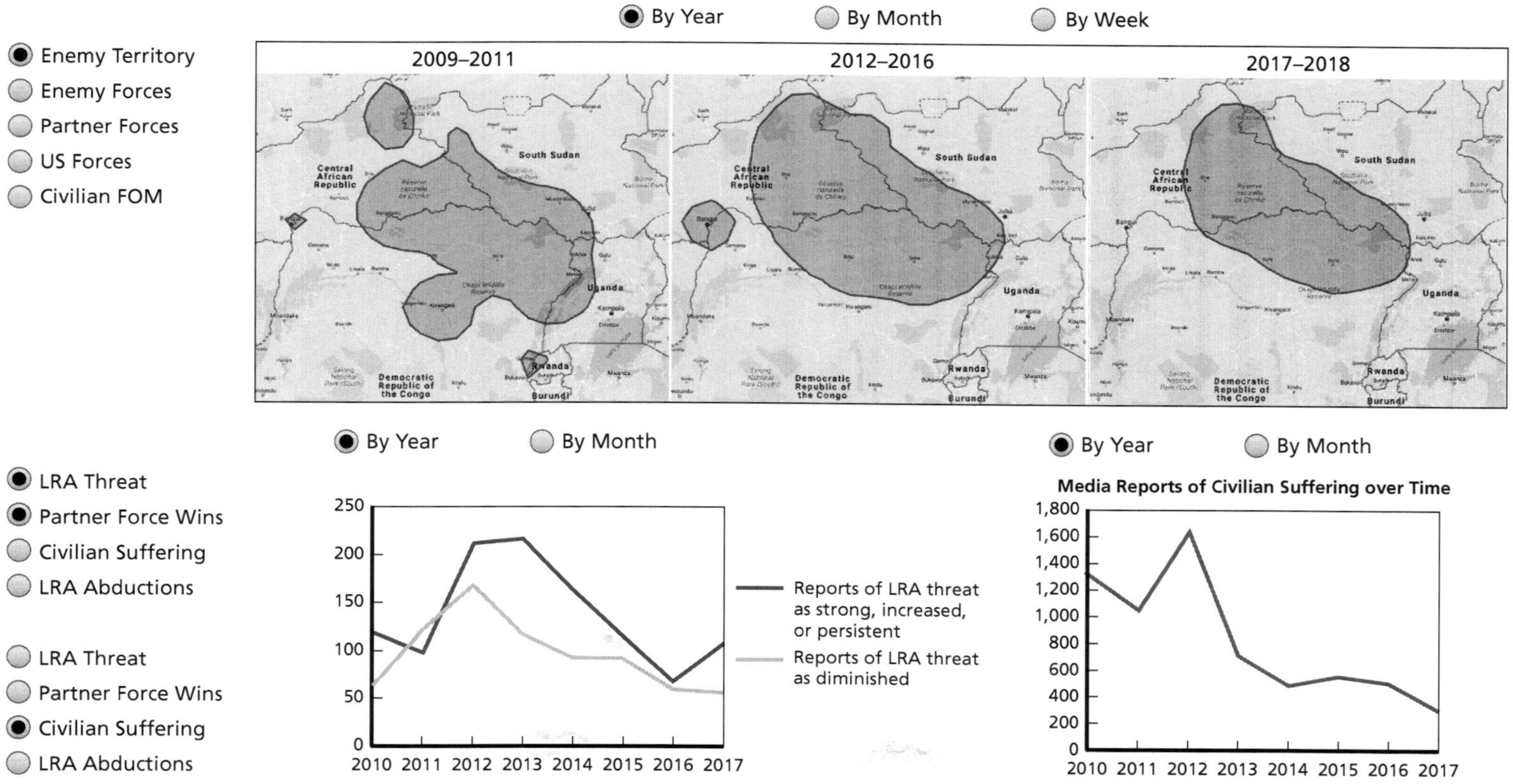

ing operational tracker or through ML-enabled extraction of data from SITREPs, the team could create a map depicting how U.S. and partner force operations were associated with changes in LRA territory over time, building from Figure 4.1 but showing change instead of static territorial control, and also depicting U.S. and partner force operational locations and intensity of operations over time.

This commander's dashboard might also be augmented with selected output from unsupervised ML analytics, focused on interrogating the data to discover unexpected relationships between and among the activities and MOEs of a campaign.[19] These analytics could be used to explore, as examples, whether activities are having delayed effects, by automatically looking for relationships between historical activities and current conditions (e.g., a reduction in violence following an operation is followed by a subsequent spike in violence); whether activities in one area is affecting conditions in another area (e.g., a reduction in violence in one area following an operation is associated with increased violence in neighboring areas); or whether MOEs are related in unexpected ways, by examining relationships among them (e.g., changes in enemy territory are associated with changes in enemy capabilities). Integrating these unsupervised ML analytics into the assessment process would likely require including a relatively sophisticated computer programmer on the assessment team, though future research might identify some "off the shelf" tools that could be used for this purpose, as described in the next chapter.

---

[19] We thank Colonel (USA) Brian Rauen for this excellent suggestion.

CHAPTER FIVE

Recommendations and Next Steps

ML can provide a powerful tool for supporting operation assessment. In the previous chapters, we presented a prototype "minimum viable" process for how an assessment team can use SML to integrate insights from information-rich—though otherwise unstructured—data sources into the assessment process. In this chapter, we discuss immediate next steps for testing and deploying this analytic process, as well as the future of ML and how it might be leveraged for assessment ten or more years in the future.

First, we provide five recommendations for how DoD can prepare to leverage ML for assessment. The first is to validate our SML approach using a controlled exercise, as this will provide a better understanding of the strengths and limitations of this approach and allow a tweaking of the proposed methods. The second is to explore how unsupervised ML techniques can be used to discover unanticipated patterns and relationships among the activities and intended effects of an operation. The third and fourth recommendations describe some modest changes to existing systems and processes that could dramatically increase the value of intelligence and operational data; these modifications are also anticipated to have other derivative benefits. The final recommendation is to expand assessment-related professional military education opportunities, expanding the ability of assessment teams and that of headquarters staff to leverage insights from assessment.

We then turn to a discussion of the likely future of ML, characterizing both the relevant capabilities and some of the risks posed by emerging approaches.

Recommendations

Our research identified five recommendations that would enhance the value of ML-based assessment approaches:

Recommendation 1: Validate the Supervised Machine Learning Approach in a Controlled Exercise

The approach described in this report was deliberately designed to be as easy as possible for an assessment team to replicate. The computer software that we used for the example in Chapter Four (see Appendixes A and B for details) was selected in large part because it is both free and already approved for use on many military networks—thus, an assessment team need only download the software and necessary analytical packages to begin incorporating SML into their assessment efforts. And the analysis in Chapter Four was implemented by a two-man team, each contributing about three days of effort once the data and analytical packages were prepared.

Most assessment teams, if provided some modest specialized premission training and augmented by an analyst with a computer programming background, should be able to leverage this SML approach to support their efforts. This premission training would likely include a general discussion of the principles of assessment and value of ML, and it should also provide an opportunity to develop and implement a mock assessment using our SML approach. The additional analyst—ideally someone familiar with the relevant computer platforms (e.g., Python, R)—would be responsible for the technical aspects of implementing SML (e.g., preparing the data, managing the SML tools), though this analyst would likely also prove useful across the broad array of analyses performed by the assessment team.[1]

However, though we have tested our approach with an actual operation, we did not attempt to do this in real time or with a team of military analysts. Thus, we recommend validating the requirements—e.g., premission training, staffing, availability of analytical tools, necessary data access—using a live exercise, but under controlled conditions.

Recommendation 2: Explore How Unsupervised Machine Learning Can Be Used to Inform Assessment and Provide a Continuous Data Stream

From the perspective of assessment, unsupervised ML can provide a tool for discovering unanticipated patterns between and among a campaign's activities and the operational objectives of those activities (as measured by the assessment team's MOEs). This can facilitate the discovery of unexpected or secondary effects of activities, or unanticipated relationships between enemy-centric, population-centric, and partner-centric objectives. Integrating these tools into the commander's dashboard could therefore provide the commander a new tool to guide decisionmaking.

Given our mandate to explore how ML could be used to leverage existing data for assessment, we did not examine in detail how these unsupervised ML tools might be used to derive new insights and support decisionmaking. We also have not assessed

[1] The analysts that the Center for Army Analysis deploys to support operational units typically have these capabilities, though many enlisted personnel, junior officers, and civilian analysts likely also have the requisite background to learn how to do this. Experience with existing intelligence collaboration platforms would be a plus.

the practicality of integrating these tools into the assessment process, and the types of analytic expertise necessary to generate useful insights from these tools. Exploring both is critical to leveraging unsupervised ML for assessment, and worthy of further investigation.

Recommendation 3: Implement Modest Standardization of Operational Reporting

SITREPs and other operational reporting are often the single best historical data source on partner force activities and capabilities. They can also provide a detailed historical record of the activities of U.S. forces, intended and actual.

These characteristics make operational reporting a potentially powerful tool for the assessment team—particularly if SML (or analogous) tools are available to rapidly extract assessment-relevant information—but this reporting is often inconsistent in quality and detail. As an important example, discussions of past activities are typically less detailed than discussions of planned activities.[2]

We recommend a modest standardization of this operational reporting: Require (simply) that tactical elements discuss their partner force's activities and capabilities on an ongoing basis. There will be a temptation by the assessment team to provide detailed proscriptions on language to use, specific fields of information to complete, and other information to satisfy campaign-relevant MOEs. However, this will increase the reporting burden for tactical elements without necessarily improving either the accuracy of reporting or facilitating assessment.[3] With the availability of SML tools to rapidly extract relevant information, simply requiring tactical elements to describe recently completed activities will substantially improve the value of this reporting.

Recommendation 4: Improve Archiving, Discovery, and Extraction of Historical Intelligence and Operational Reporting

Accessing and extracting the relevant intelligence and operational reporting necessary for the approaches described in this report is not easy with existing systems. Our fourth recommendation contains the following two sub-recommendations about how these data can more easily be leveraged for assessment:

For intelligence reporting, we recommend a modest modification to existing knowledge management systems within the intelligence enterprise (e.g., Multimedia

---

[2] The ideal daily SITREP, particularly for U.S. forces working with a partner force, includes a detailed list of recent and planned activities (see "Appendix C: Situation Report" in Center for Army Lessons Learned, *Theater Security Cooperation: The Military Engagement Team, Lessons and Best Practices*, 2017).

[3] In some cases, the assessment team may be able to develop a set of information requirements that would provide all the information needed for certain MOEs. If this is possible, it should be communicated to the commander and made into a commander's critical information requirement. However, this should be done with caution, as operators may often interpret the guidance for these type of unusual requirements differently, leading to less-than-accurate results, and these types of requirements can also affect the behavior of operations, potentially in perverse ways.

Message Manager). While these systems are effective at discovering relevant intelligence, they are typically designed to allow the analyst to review a single intelligence report at a time—thus, they do not have the capability to extract the full text of all (potentially) relevant reporting into Excel or other comparable file format for subsequent analysis.[4] We recommend that these systems be augmented with a capability to rapidly extract longitudinal data (e.g., three or more months of data) for a given set of search parameters (e.g., geographic, Boolean logic) that would contain the full text of the report alongside some basic descriptive information about the data (date, source, etc.).[5] This approach, while simple, is also anticipated to be the most flexible, as it will facilitate interoperability with the specialized software frequently required for ML and other leading-edge technical capabilities.

For operational reporting, we recommend that operational-level headquarters ensure that SITREPs from all subordinate headquarters are systematically archived. Frequently distributed via email, the operational information included in these SITREPs is rarely archived in a single location to allow later analysis. Some existing systems, described in the classified appendix in greater detail, provide examples of how this might be implemented.[6]

Recommendation 5: Expand Assessment-Specific Discussions Required in Professional Military Education

Despite the acknowledged importance of assessment to military operations, there are major gaps in the assessment-related education and training provided to the joint force.[7] As a consequence, assessment teams often struggle to provide the information necessary to guide effective decisionmaking,[8] and other staff at the operational-level headquarters often lack the experience and training to effectively integrate assessments—even when it is accurate, relevant, and timely—into operations and planning.[9]

---

[4] Some knowledge management platforms have versions of this capability, as discussed in the classified appendix in greater detail.

[5] Some knowledge management platforms have versions of this capability, as discussed in the classified appendix in greater detail. An alternative solution would be to integrate ML tools into these knowledge management platforms themselves; while this would likely create ancillary benefits for intelligence analysts, by allowing for content-based and not only keyword-based searches, it would require significant time and resources to implement within these existing platforms.

[6] As an example, some U.S. SOF formations have archived SITREPs in easy-to-analyze database formats (e.g., Special Operations Command Africa), though in many cases these SITREPs are only available in email format.

[7] Lynette M. B. Arnhart and Marvin L. King, "Are We There Yet? Implementing Best Practices in Assessments," *Military Review*, May–June 2018.

[8] Arnhart and King, 2018.

[9] Assessment—when properly executed and integrated into command decisionmaking—should guide both operations and planning (U.S. Army War College, 2016; Air Land Sea Application Center, 2015 Army Doctrine Reference Publication 5-0, 2012).

Given the increased complexity of assessments leveraging SML methods, these difficulties are likely to be exacerbated for headquarters using these methods. Fully exploiting intelligence, operational, and ambient data will require staff from across the headquarters (e.g., G-2, G-3) to be willing and able to support the assessment team, and the assessment team must have the appropriate background to understand and effectively analyze these diverse data.[10] Effectively leveraging the real-time analysis possible with SML methods will also require similarly flexible and prepared personnel.

We recommend—echoing the recommendation of others[11]—expanding the assessment-related material offered to both officers and enlisted personnel in professional military education. This will enable operational-level headquarters to more effectively integrate SML and other emerging technologies into assessment and will, more broadly, prepare the joint force for the gradually increasing assessment requirements for military operations.[12]

Future Directions in Machine Learning and Assessment

Advances in Supervised Machine Learning–Based Text Classification: Improvements in Performance

Natural language processing (NLP) refers to a variety of analytic approaches designed to allow machines to map and analyze the features and structures of human language. The specific SML approach used in this report is a text classifier, which is one type of NLP. Newer text classifier algorithms will likely make approaches such as the one described in this report more accurate, and potentially easier to implement and troubleshoot.[13] These new approaches, which involve a mixture of supervised and unsupervised learning, have shown increased performance in difficult tasks for humans

[10] This required collaboration may have the derivative benefit of increasing overall coordination across the command, as has been observed in other assessments with a similar requirement (e.g., Schroden et al., 2013, p. 18).

[11] Arnhart and King, 2018; Robinson et al., 2017; Schroden, 2011.

[12] This trend is demonstrated by the 2017 National Defense Authorization Act requirement that DoD "develop and maintain an assessment, monitoring, and evaluation framework for security cooperation with foreign countries to ensure accountability and foster implementation of best practices" (Section 1205 in Public Law 114-328, National Defense Authorization Act for Fiscal Year 2017, December 23, 2016). As a result, DoD issued Department of Defense Instruction 5132.14, which made policy to conduct "independent and rigorous evaluations of significant security cooperation initiatives to examine their relevance, effectiveness, and sustainability" (Department of Defense Instruction 5132.14, *Assessment, Monitoring, and Evaluation Policy for the Security Cooperation Enterprise*, Washington, D.C.: Office of the Under Secretary of Defense for Policy, January 13, 2017).

[13] These approaches are often more computationally demanding, as they employ deep learning or "ensemble" approaches, but should not be much more difficult to implement. See pp. 6–14 in Laura Anderlucci, Lucia Guastadisegni, and Cinzia Viroli, "Classifying Textual Data: Shallow, Deep and Ensemble Methods," *arXiv*, February 18, 2019.

that involve low signal-to-noise ratio, such as predicting suicide risk from social media content.[14]

Stance and Sentiment Analysis: Using Known Features of Language

One growing field of NLP is "stance and sentiment" analysis, which relies on reference libraries of English and other languages to label rhetorical position ("stance"), emotional tone ("sentiment"), and other features of text content. For example, such an approach can be used to automatically detect the degree to which the author of a tweet supports a politician or leader of social movements, or the degree to which they are expressing anger or other emotions regarding an ongoing war. Such analytic techniques can give the assessment team additional insights into nuances and subtleties in the way that the local population or enemy is feeling and expressing themselves. Increasingly, such analysis is available across languages[15]—including Russian,[16] Chinese,[17] and Arabic.[18]

Anomaly Detection: Indicators and Warnings

ML—particularly unsupervised learning—has proven to be useful for anomaly detection;[19] for example, searching for patterns indicating the presence of attempted cyber attack or credit card fraud.[20] Integrating anomaly detection into data analytics processes for assessment could help alert the assessment team to cases in which the enemy may have changed tactics or the local population may be dramatically shifting in its opinions. In additional, a combination of supervised and unsupervised tech-

---

[14] Glen Coppersmith, Ryan Leary, Patrick Crutchley, and Alex Fine, "Natural Language Processing of Social Media as Screening for Suicide Risk," *Biomedical Informatics Insights*, Vol. 10, 2018.

[15] For example, see Braja Gopal Patra, Dipankar Das, and Amitava Das, "Sentiment Analysis of Code-Mixed Indian Languages: An Overview of SAIL_Code-Mixed Shared Task@ ICON-2017," *arXiv*, March 18, 2018. Also see, Dilek Küçük and Fazli Can, "A Tweet Dataset Annotated for Named Entity Recognition and Stance Detection," *arXiv*, January 16, 2019.

[16] Nikita Lozhnikov, Leon Derczynski, and Manuel Mazzara, "Stance Prediction for Russian: Data and Analysis," *International Conference in Software Engineering for Defence Applications*, 2018, pp. 176–186.

[17] Da Pan, Meishan Zhang, and Guohong Fu, "Chinese Sentiment Analysis Exploiting Heterogeneous Segmentations," *CCL*, 2016, pp. 387–398.

[18] Ayman El-Kilany, Amr Azzam, and Samhaa R. El-Beltagy, "Using Deep Neural Networks for Extracting Sentiment Targets in Arabic Tweets," in Khaled Shaalan, Aboul Ella Hassanien, and Mohammed F. Tolba, eds., *Intelligent Natural Language Processing: Trends and Applications*, New York: Springer, 2018, pp. 3–15.

[19] Shikha Agrawal and Jitendra Agrawal, "Survey on Anomaly Detection Using Data Mining Techniques," *Procedia Computer Science*, Vol. 60, 2015, pp. 708–713.

[20] Mohiuddin Ahmed, Abdun Naser Mahmood, and Jiankun Hu, "A Survey of Network Anomaly Detection Techniques," *Journal of Network Computer Applications*, Vol. 60, 2016, pp. 19–31; Nick F. Ryman-Tubb, Paul Krause, and Wolfgang Garn, "How Artificial Intelligence and Machine Learning Research Impacts Payment Card Fraud Detection: A Survey and Industry Benchmark," *Engineering Applications of Artificial Intelligence*, Vol. 76, 2018, pp. 130–157.

niques may be used to detect fake or "bot" social media accounts or other artificial interlocutors in assessment-relevant data streams in order to either remove them from analysis or analyze their behavior.[21]

Caveats and Risks

A central observation from our research is that ML tools do not replace human analysts, but instead act as a "force multiplier" by allowing analysts to automate repetitive tasks. Though emergent technologies will likely make these tools more effective and flexible, neither near-term nor mid-term developments in ML are likely to make human analysts obsolete, for either assessment or other DoD applications. Indeed, the state of ML—and the larger field of research in artificial intelligence—is still very far from the ability to design the assessment or coding process and analyze or even interpret data on its own.[22]

In addition to emerging capabilities that will support assessment, there is now evidence that it is possible to deliberately mislead even the most advanced ML algorithms.[23] As an example, a skilled adversary might be able to manipulate surface features (or even deliberately edit publicly available imagery) so that ML tools "see" a tank where there is not one. Though the implications of this ability to "fool" or "hack" ML algorithms has not been fully explored, it creates a potential vulnerability for assessments that are heavily reliant on ML. Assessments are designed to guide commander decisionmaking, and a skilled adversary might be able to influence the ML-derived insights included in this assessment. Analysts are undoubtedly cognizant of this risk for other forms of insight (e.g., HUMINT) and should similarly not rely on a single type of ML-derived insight.

---

[21] Alex Hai Wang, "Detecting Spam Bots in Online Social Networking Sites: A Machine Learning Approach," *24th Annual IFIP WG 11.3 Working Conference on Data and Applications Security and Privacy (DBSEC)*, Rome, Italy, June 2010, pp. 335–342.

[22] Indeed, while artificial intelligence has been used for some exploratory correlation analysis, applications for causal inference are still very much in development (e.g., Judea Pearl, "The Seven Tools of Causal Inference with Reflections on Machine Learning," *Communications of Association for Computing Machinery*, 2018). For ML to have this capability, it would need to be able to mimic the cognitive and logical processes of human analysts—this type of capability, commonly referred to as "general artificial intelligence," is far beyond our current capabilities (for a discussion, see Mark Rosa and Jan Feyereisl, "A Framework for Searching for General Artificial Intelligence," *arXiv*, 2016).

[23] Adam Geitgey, "Machine Learning Is Fun Part 8: How to Intentionally Trick Neural Networks: A Look into the Future of Hacking," *Medium*, August 16, 2017.

APPENDIX A

Support Vector Machine

The specific ML algorithm that we use in this report is a "support vector machine" (SVM),[1] a statistical approach to ML that has been widely used for analysis of text and imagery. SVMs perform well in analysis of complex text and require less training data than more-complex models, such as deep neural networks.[2] SVMs have been demonstrated to be a practical mid-point between algorithmic complexity and accuracy.[3]

One common approach for using SVM in text classification (the application of focus in this report) is to compute individual histograms of word frequencies for each report, the so-called "bag-of-words" feature, and then use these histograms as inputs to the SVM. For this report, we selected a Gaussian kernel SVM (one type of SVM classifier) with the bag-of-words feature, as this combination is known to perform well on similar text classification efforts.[4] In our case, this combination allowed the assessment team to accurately extrapolate from a relatively small training dataset (e.g., approximately 300 hand-coded reports) to much larger datasets (tens of thousands of reports).

This SVM text classifier can be implemented in a variety of widely used computer programming suites, including several that are pre-authorized for use on DoD classified systems (e.g., R, Python),[5] using freely available tools.[6] Implementation is relatively straightforward, and an analyst with a moderate computer programming background

[1] See p. 47 in Marcin Michał Mirończuk and Jarosław Protasiewicz, "A Recent Overview of the State-of-the-Art Elements of Text Classification," *Expert Systems with Applications*, Vol. 106, 2018, pp. 36–54.

[2] See Bala Deshpande, "When Do Support Vector Machines Trump Other Classification Methods," Simafore, blog post, January 28, 2013.

[3] For example, Mikhail Zaydman, *Tweeting About Mental Health: Big Data Text Analysis of Twitter for Public Policy*, Santa Monica, Calif.: RAND Corporation, RGSD-391, 2017.

[4] Mirończuk and Protasiewicz, 2018.

[5] These two computer programs are "R" and "Python." Python is currently being used by DoD intelligence professionals within the USASOC enterprise and R is the preferred computer program for many civilian analysts across the DoD (e.g., the Defense Advanced Research Projects Agency).

[6] For R, which we used in this report, we use the SVM tool developed and maintained by the Quanted Initiative (https://quanteda.io/). For Python, the authors have used the scikit-learn package for implementing a SVM (https://scikit-learn.org/stable/).

should be able to adapt these algorithms to their needs.[7] We demonstrate the relative simplicity of this tool in our example in Chapter Four and in the sample computer code provided in Appendix B.

There are a range of free and publicly available text classifiers at a similar level of computational complexity to SVM, including Naïve-Bayes, k-Nearest Neighbor, Component-wise classifiers, Linear Discriminant Analysis, and Decision Trees/Random Forests. Each of these has computational costs or other vulnerabilities,[8] which make them less well suited for the types of text classification tasks likely to be found in operation assessment.[9] Newer and more computationally demanding text classification approaches using deep learning or "ensemble" (multiple classifiers acting as a "committee") approaches may produce higher performance, but they are considerably more difficult to implement.[10]

---

[7] Though relatively easily to implement, the integration of the SVM into the assessment team's workflow will require a programmer with experience in R or Python—this is necessary to allow the assessment team to make minor modifications to the computer code, analogous to any type of quantitative analysis.

[8] See pp. 3–5 in Anderlucci, Guastadisegni, and Viroli, 2019.

[9] We (informally) tested the Naïve-Bayes text classifier in Quanteda and did not notice significant departures from results using the SVM approach.

[10] See pp. 6–14 in Anderlucci, Guastadisegni, and Viroli, 2019.

APPENDIX B

Code for Example in Chapter Four

This appendix provides the computer code used for producing the analysis presented in Chapter Four. We have included this appendix because we want to demonstrate that ML algorithms are not overly difficult for an assessment team to integrate into their efforts. Indeed, the computer code presented here is sufficiently simple that it can be easily adapted by a moderately experienced computer programmer to almost any operational context. Further, the computer program that we have used—R—is free, publicly available, and pre-accredited for DoD classified computer systems. The research team has developed analogous code for Python, another widely-used and free programming language, which is available upon request from the corresponding author.

This computer code is designed to prepare, process, and analyze unstructured text data available in Excel format following the approach outlined in Chapter Two. For the example in this appendix, we used extracts of newspaper data available from LexisNexis that mention "LRA" or "Lord's Resistance Army." A sample of the raw data used for this analysis presented in Figure B.1—the data that is the focus of our ML tools is that provided in the "Text" column.

Figure B.1
Example of Raw Data Used for Analysis

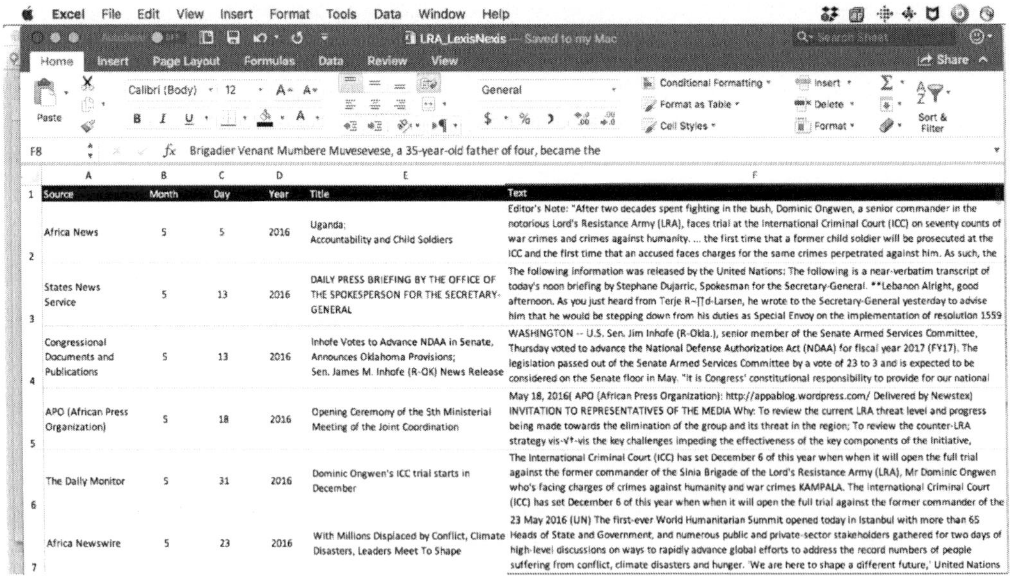

Source	Month	Day	Year	Title	Text
Africa News	5	5	2016	Uganda: Accountability and Child Soldiers	Editor's Note: "After two decades spent fighting in the bush, Dominic Ongwen, a senior commander in the notorious Lord's Resistance Army (LRA), faces trial at the International Criminal Court (ICC) on seventy counts of war crimes and crimes against humanity. ... the first time that a former child soldier will be prosecuted at the ICC and the first time that an accused faces charges for the same crimes perpetrated against him. As such, the
States News Service	5	13	2016	DAILY PRESS BRIEFING BY THE OFFICE OF THE SPOKESPERSON FOR THE SECRETARY-GENERAL	The following information was released by the United Nations: The following is a near-verbatim transcript of today's noon briefing by Stephane Dujarric, Spokesman for the Secretary-General. **Lebanon Alright, good afternoon. As you just heard from Terje R~∏d-Larsen, he wrote to the Secretary-General yesterday to advise him that he would be stepping down from his duties as Special Envoy on the implementation of resolution 1559
Congressional Documents and Publications	5	13	2016	Inhofe Votes to Advance NDAA in Senate, Announces Oklahoma Provisions; Sen. James M. Inhofe (R-OK) News Release	WASHINGTON -- U.S. Sen. Jim Inhofe (R-Okla.), senior member of the Senate Armed Services Committee, Thursday voted to advance the National Defense Authorization Act (NDAA) for fiscal year 2017 (FY17). The legislation passed out of the Senate Armed Services Committee by a vote of 23 to 3 and is expected to be considered on the Senate floor in May. "It is Congress' constitutional responsibility to provide for our national
APO (African Press Organization)	5	18	2016	Opening Ceremony of the 5th Ministerial Meeting of the Joint Coordination	May 18, 2016(APO (African Press Organization): http://appablog.wordpress.com/ Delivered by Newstex) INVITATION TO REPRESENTATIVES OF THE MEDIA Why: To review the current LRA threat level and progress being made towards the elimination of the group and its threat in the region; To review the counter-LRA strategy vis-√†-vis the key challenges impeding the effectiveness of the key components of the Initiative,
The Daily Monitor	5	31	2016	Dominic Ongwen's ICC trial starts in December	The International Criminal Court (ICC) has set December 6 of this year when when it will open the full trial against the former commander of the Sinia Brigade of the Lord's Resistance Army (LRA), Mr Dominic Ongwen who's facing charges of crimes against humanity and war crimes KAMPALA. The International Criminal Court (ICC) has set December 6 of this year when when it will open the full trial against the former commander of the
Africa Newswire	5	23	2016	With Millions Displaced by Conflict, Climate Disasters, Leaders Meet To Shape	23 May 2016 (UN) The first-ever World Humanitarian Summit opened today in Istanbul with more than 65 Heads of State and Government, and numerous public and private-sector stakeholders gathered for two days of high-level discussions on ways to rapidly advance global efforts to address the record numbers of people suffering from conflict, climate disasters and hunger. 'We are here to shape a different future,' United Nations

Figure B.2 provides computer code for the first step of our five-step ML analytical workflow (this is the "Curate Data" step). This code identifies specific sentences that mention keywords of interest—in our case "LRA" or "Lord's Resistance Army"—and then extracts that sentence and the sentences neighboring it. These clusters of three sentences become the basis for the training data, which can then be used to develop the coding criteria (Step 2 of the workflow) and build the training data.

Figure B.3 illustrates the training data that is the output of Step 2 ("Develop Coding Criteria") and Step 3 ("Construct Training Data"). In this example, we have coded each three-cluster sentence produced by the data curation process along six criteria. These are the six criteria underlying the analysis in Figure 4.5, and ordered as follows: (1) "PF_score," report mentions a partner force success, (2) "PF_US," report mentions the United States working with a partner force, (3) "PF_neg," report mentions a U.S. partner force in a negative light, (4) "Threat_up", report mentions LRA threat increased, (5) "Threat_down," whether LRA threat has decreased, and (6) "Citizen_impact," whether citizens are impacted by LRA violence.

Figure B.2
Prepare Unstructured Data for Analysis

```
# 1: Load necessary R packages #
library(tidyr)

# 2: Load and clean-up the data #
news <- read.csv("Lexis-Nexis_LRA.csv")
news$ID <- 1:nrow(news)
news$text <- iconv(substr(news$text,1,10000),to="ASCII")
news$text <- gsub("\n", " ", news$text, fixed = TRUE)
news$text <- gsub("([.])|[[:punct:]]", "\\1", news$text)
news$text <- gsub("U\\.S\\.", "US", news$text)

# 3: Break the long narrative discussions into individual sentences - keep first 100 #
sentences <- strsplit(as.character(news$text), "\\.")
sentences_data <- plyr::ldply(sentences, rbind)
sentences_data <- sentences_data[,1:100]

# 4: Identify all sentences that mention LRA #
for (i in 1:ncol(sentences_data)) { colnames(sentences_data)[i]<- paste("col",i,sep="")
    }
sentences_data <- cbind(news$ID,sentences_data)
colnames(sentences_data)[1]<- "ID"
sentences_data_long <- reshape(sentences_data_transpose, varying= c(2:ncol(sentences_
    data_transpose)), direction="long", idvar="ID", sep="",timevar="order")
sentences_data_long <- sentences_data_long[complete.cases(sentences_data_long),]
LRA <- (regexpr("LRA",sentences_data_long$col)+regexpr("Lords Resistance Army",
    sentences_data_long$col))>0

# 5: Remove duplicated sentences, selecting first use of each sentence #
sentences_data_LRA <- sentences_data_LRA[nchar(as.character(sentences_data_
    LRA$col))<1000,]
sentences_data_LRA$duplicates <- substr(gsub("[^[:alnum:]]","",sentences_data_
    LRA$col),1,100)
news_dates <- news[,c("ID","year","month","day")]
test <- merge(x=sentences_data_LRA,y=news_dates,by="ID")
test <- test[order(test$duplicates,test$year,test$month,test$day),]
res <- test[!duplicated(test$duplicates),]
data_prepared <- res[,c("ID","col")]

# 6: Build clusters of 3 sentences around LRA-mentioning sentence #
master_data <- merge(x=data_prepared,y=news,by="ID")
for (i in 1:nrow(master_data)) {
    position <- regexpr(as.character(master_data$col[i]),as.character(master_
        data$text[i]))
    previous_text <- substr(as.character(master_data$text[i]),1,position)
    sentences_previous <- strsplit(previous_text, "\\.")
    master_data$col_previous[i] <- as.character(lapply(sentences_previous,
        dplyr::nth,-2))
    following_text <- substr(as.character(master_data$text[i]), position+nchar(as.
        character(master_data$col[i])),nchar(as.character(master_data$text[i])))
    sentences_following <- strsplit(following_text, "\\.")
    master_data$col_following[i] <- as.character(lapply(sentences_following,
        dplyr::nth,2))
}
master_data$three_sentences <- paste(master_data$col_previous, master_data$col,
    master_data$col_following, sep=".")

# 7: Export the data #
drop <- c("text","col_previous","col_following")
export_data <- master_data[,!(names(master_data) %in% drop )]
write.csv(export_data,file = "LRA_Prepared_Data.csv")
```

Figure B.3
Sample Training Data

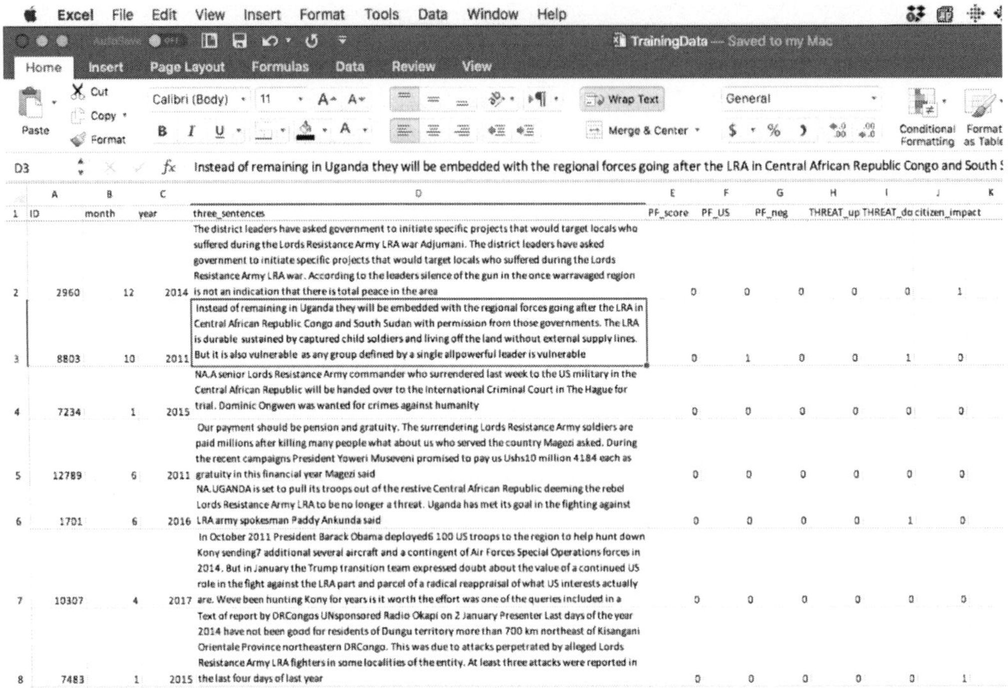

	ID	month	year	three_sentences	PF_score	PF_US	PF_neg	THREAT_up	THREAT_do	citizen_impact
2	2960	12	2014	The district leaders have asked government to initiate specific projects that would target locals who suffered during the Lords Resistance Army LRA war Adjumani. The district leaders have asked government to initiate specific projects that would target locals who suffered during the Lords Resistance Army LRA war. According to the leaders silence of the gun in the once warravaged region is not an indication that there is total peace in the area	0	0	0	0	0	1
3	8803	10	2011	Instead of remaining in Uganda they will be embedded with the regional forces going after the LRA in Central African Republic Congo and South Sudan with permission from those governments. The LRA is durable sustained by captured child soldiers and living off the land without external supply lines. But it is also vulnerable as any group defined by a single allpowerful leader is vulnerable	0	1	0	0	1	0
4	7234	1	2015	NA.A senior Lords Resistance Army commander who surrendered last week to the US military in the Central African Republic will be handed over to the International Criminal Court in The Hague for trial. Dominic Ongwen was wanted for crimes against humanity	0	0	0	0	0	0
5	12789	6	2011	Our payment should be pension and gratuity. The surrendering Lords Resistance Army soldiers are paid millions after killing many people what about us who served the country Magezi asked. During the recent campaigns President Yoweri Museveni promised to pay us Ushs10 million 4184 each as gratuity in this financial year Magezi said	0	0	0	0	0	0
6	1701	6	2016	NA.UGANDA is set to pull its troops out of the restive Central African Republic deeming the rebel Lords Resistance Army LRA to be no longer a threat. Uganda has met its goal in the fighting against LRA army spokesman Paddy Ankunda said	0	0	0	0	1	0
7	10307	4	2017	In October 2011 President Barack Obama deployed6 100 US troops to the region to help hunt down Kony sending7 additional several aircraft and a contingent of Air Forces Special Operations forces in 2014. But in January the Trump transition team expressed doubt about the value of a continued US role in the fight against the LRA part and parcel of a radical reappraisal of what US interests actually are. Weve been hunting Kony for years is it worth the effort was one of the queries included in a	0	0	0	0	0	0
8	7483	1	2015	Text of report by DRCongos UNsponsored Radio Okapi on 2 January Presenter Last days of the year 2014 have not been good for residents of Dungu territory more than 700 km northeast of Kisangani Orientale Province northeastern DRCongo. This was due to attacks perpetrated by alleged Lords Resistance Army LRA fighters in some localities of the entity. At least three attacks were reported in the last four days of last year	0	0	0	0	0	1

Figure B.4 illustrates the computer code used to implement the SVM algorithm based on the training data, which is Step 4 in our five-step process. This computer code loads the necessary analytical packages and the training data developed by the assessment team, builds a corpus of text for the three-sentence blocks that we built previously (see previous paragraph and Figure B.2), and then implements the ML algorithm. The intuition for this ML algorithm is that it will be looking for text that is "overrepresented" in reports that our training data identified as relevant, and then searches for other reports that have similar text.

Figure B.4
Implementation of Machine Learning Algorithm

```
# 1: Load necessary R packages #
library(quanteda)
library(kernlab)

# 2: Load and prepare training data #
kdata <- read.csv("TrainingData.csv")
list_variables <- c("THREAT_up","THREAT_down","citizen_impact","PF_score","PF_
    US","PF_neg")
kdata <- kdata[c("ID","month","year","three_sentences",list_variables)]
kdata$cleaned.body<-tolower(kdata$three_sentences)
kdata$cleaned.body <- gsub("[[:punct:]]", "", kdata$cleaned.body)
training.row <- 300

# 3: Establish the corpus of text from the full set of data #
kdtm<-dfm(kdata$cleaned.body,remove=stopwords("English"))
kwf<-docfreq(kdtm)
words.to.keep<-names(sort(kwf,decreasing=T)[1:200])
kdtm2<-kdtm[,words.to.keep]
k.freq<-dfm_tfidf(kdtm2)
new.data2<-k.freq[1:training.row,]
new.data2<-as.data.frame(as.matrix(k.freq[1:training.row,],sparse=F))

# 4: Implement the ML algorithm for each of the six variables in the training data
    #
for (i in 1:length(list_variables)) {
    new.data2_outcome <- cbind(new.data2,kdata[1:training.row,colnames(kdata)==list_
        variables[i]])
    colnames(new.data2_outcome)[ncol(new.data2_outcome)]<-"outcome"
    new.data2_outcome$outcome[is.na(new.data2_outcome$outcome)] <- "0"
    new.data2_outcome$outcome <- as.factor(as.character(new.data2_outcome$outcome))
    # Note: This following three lines implements the SVM algorithm #
    refind.outcome<-ksvm(outcome~.,data=new.data2_outcome,prob.model=T,kernel="rbfd
        ot",C=50,cross=10)
    new.data3<-k.freq[(training.row+1):nrow(k.freq),]
    predict.values<-predict(refind.outcome,new.data3,type="probabilities")
    # Note: The remainder of the code saves the predicted probabilities #
    kdata$predicted.values<-c()
    kdata$predicted.values[(training.row+1):nrow(kdata)]<-as.character(predict.
        values[,2]
    names(kdata)[names(kdata) == "predicted.values"] <- paste("predicted.
        values_",list_variables[i],sep="")
}
kdata <- kdata[ , !(names(kdata) %in% "cleaned.body")]
write.csv(kdata,file="Output.csv")
```

A key component of implementing the ML algorithm is the calibration process (Step 4 of the five-step process), which in the case of SVM is the modification of a number of "hyperparameters." First among these is the type of kernel function (polynomial and associated degree versus Gaussian and associated width parameter) that the ML algorithm is using. This affects the type of boundary used to distinguish between cases and noncases, as well as the degree to which the SVM uses complex curvature to fit the boundary between cases and noncases in the data. If using radial basis kernels (typically Gaussians), excessive Gaussian width can lead to model overfitting. Kernel type and associated parameters can be adjusted by altering the bold text in the following line of code:

```
refind.outcome<ksvm(outcome~.,data=new.data2_outcome,prob.
model=T,kernel="rbfdot",C=50,cross=10.
```

Other hyperparameters that can be tuned include "*C*," a regularization parameter that determines the cost of misclassifying samples in the training set. Large values of *C* will lead to a low error rate on the training data but may lead to overfitting. Conversely, low values of *C* will lead to a higher error rate on the training data but with a lesser risk of overfitting. The tuning of *C* generally must be done in an iterative process. Values for *C* can be adjusted by altering the bold text in the following line of code:

```
refind.outcome<ksvm(outcome~.,data=new.data2_outcome,prob.
model=T,kernel="rbfdot",C=50,cross=10.
```

Finally, the user may alter the number of cross-validation folds by altering the bold text in the following line of code. The appropriate number of cross-validation folds is also dependent on the data at hand and must be tuned through trial-and-error, although ten-fold cross-validation is a common default option:

```
refind.outcome<ksvm(outcome~.,data=new.data2_outcome,prob.
model=T,kernel="rbfdot",C=50,cross=10.
```

The output from the computer code in Figure B.4 can then be used directly to support longitudinal analysis of MOEs—as demonstrated in Figures 4.1, 4.2, and 4.4—using standard analytical tools in Excel. However, ML-based analysis of data that includes geospatial information (e.g., intelligence reporting) can also be used to create geographical overlays of enemy freedom of movement, partner activities, etc. One simple approach for doing this, which is the tool that we used in creating Figure 4.3, is provided in Figure B.5. This sample uses the geospatial information contained in the ACLED database to make maps of enemy freedom of movement. This code could be readily adapted to output from the SML approach and can be applied to output from intelligence, operational, and ambient data.

Figure B.5
Measuring Freedom of Movement

```
# 1: Load necessary R packages #
library(quanteda)
library(kernlab)

# 2: Load and prepare training data #
kdata <- read.csv("TrainingData.csv")
list_variables <- c("THREAT_up","THREAT_down","citizen_impact","PF_score","PF_
    US","PF_neg")
kdata <- kdata[c("ID","month","year","three_sentences",list_variables)]
kdata$cleaned.body<-tolower(kdata$three_sentences)
kdata$cleaned.body <- gsub("[[:punct:]]", "", kdata$cleaned.body)
training.row <- 300

# 3: Establish the corpus of text from the full set of data #
kdtm<-dfm(kdata$cleaned.body,remove=stopwords("English"))
kwf<-docfreq(kdtm)
words.to.keep<-names(sort(kwf,decreasing=T)[1:200])
kdtm2<-kdtm[,words.to.keep]
k.freq<-dfm_tfidf(kdtm2)
new.data2<-k.freq[1:training.row,]
new.data2<-as.data.frame(as.matrix(k.freq[1:training.row,],sparse=F))

# 4: Implement the ML algorithm for each of the six variables in the training data #
for (i in 1:length(list_variables)) {
    new.data2_outcome <- cbind(new.data2,kdata[1:training.row,colnames(kdata)==list_
        variables[i]])
    colnames(new.data2_outcome)[ncol(new.data2_outcome)]<-"outcome"
    new.data2_outcome$outcome[is.na(new.data2_outcome$outcome)] <- "0"
    new.data2_outcome$outcome <- as.factor(as.character(new.data2_outcome$outcome))
    # Note: This following three lines implements the SVM algorithm #
    refind.outcome<-ksvm(outcome~.,data=new.data2_outcome,prob.model=T,kernel="rbfd
        ot",C=50,cross=10)
    new.data3<-k.freq[(training.row+1):nrow(k.freq),]
    predict.values<-predict(refind.outcome,new.data3,type="probabilities")
    # Note: The remainder of the code saves the predicted probabilities #
    kdata$predicted.values<-c()
    kdata$predicted.values[(training.row+1):nrow(kdata)]<-as.character(predict.
        values[,2]
    names(kdata)[names(kdata) == "predicted.values"] <- paste("predicted.
        values_",list_variables[i],sep="")
}
kdata <- kdata[ , !(names(kdata) %in% "cleaned.body")]
write.csv(kdata,file="Output.csv")
```

References

ACLED, "Methodology," webpage, undated. As of November 7, 2019: https://www.acleddata.com/resources/methodology/

Agrawal, Shikha, and Jitendra Agrawal, "Survey on Anomaly Detection Using Data Mining Techniques," *Procedia Computer Science*, Vol. 60, 2015, pp. 708–713.

Ahmed, Mohiuddin, Abdun Naser Mahmood, and Jiankun Hu, "A Survey of Network Anomaly Detection Techniques," *Journal of Network and Computer Applications*, Vol. 60, 2016, pp. 19–31.

Air Land Sea Application Center, *Operation Assessment: Multi-Service Tactics, Techniques, and Procedures for Operation Assessment*, ATP 5-0.3, MCRP 5-1C, NTTP 5-01.3, AFTTP 3-2.87, 2015.

Anderlucci, Laura, Lucia Guastadisegni, and Cinzia Viroli, "Classifying Textual Data: Shallow, Deep and Ensemble Methods," *arXiv*, February 18, 2019.

Anderse, Therese, Hong Xu, Cheng Cheng, and T. K. Satish Kumar, "Measuring Territorial Control in Civil Wars Using Hidden Markov Models: A Data Informatics-Based Approach," 31st Conference on Neural Information Processing Systems, 2017.

Army Doctrine Reference Publication 5-0, *The Operations Process*, Washington, D.C.: Headquarters, Department of the Army, 2012.

Army Techniques Publication 3-55.4, *Techniques for Information Collection During Operations Among Populations*, Washington, D.C.: Headquarters, Department of the Army, April 2016.

Arnhart, Lynette M. B., and Marvin L. King, "Are We There Yet? Implementing Best Practices in Assessments," *Military Review*, May–June 2018.

Axis Communications, "Solutions by Application," webpage, undated. As of November 7, 2019: https://www.axis.com/en-us/solutions-by-application

Baddorf, Zack, "Uganda Ends Its Hunt for Joseph Kony Empty-Handed," *New York Times*, April 20, 2017.

Bhaduri, Budhendra, Edward Bright, Phillip Coleman, and Marie L. Urban, "LandScan USA: A High-Resolution Geospatial and Temporal Modeling Approach for Population Distribution and Dynamics," *GeoJournal*, Vol. 69, No. 1, 2007, pp. 103–117.

Blanken, Leo J., and Jason J. Lepore, "Principals, Agents, and Assessment," in Leo J. Blanken, Hy Rothstein, and Jason J. Lepore, eds., *Assessing War: The Challenge of Measuring Success and Failure*, Washington, D.C.: Georgetown University Press, 2015.

Blinde, Loren, "DoD Announces 2nd AI Industry Day," *Intelligence Community News*, October 19, 2018. As of November 7, 2019:
https://intelligencecommunitynews.com/dod-announces-2nd-ai-industry-day/

Borra, Simone, and Agostino Di Ciacco, "Measuring the Prediction Error: A Comparison of Cross-Validation, Bootstrap and Covariance Penalty Methods," *Computational Statistics and Data Analysis*, Vol. 54, pp. 2976–2989.

Bourret, Andrew K., Joshua D. Wines, and Jason M. Mendes, "Assessing Sentiment in Conflict Zones Through Social Media," Monterey, Calif.: Naval Postgraduate School, 2016.

Bulos, Nabih, "Libya's Two Wars: One on the Battlefield and One on Facebook," *Los Angeles Times*, June 17, 2019.

Burgman, Mark A., and Julian C. Fox, "Bias in Species Range Estimates from Minimum Convex Polygons: Implications for Conservation and Options for Improved Planning," *Animal Conservation*, Vol. 6, No. 1, February 2003.

Campbell, John L., Charles Quincy, Jordan Osserman, and Ove K. Pedersen, "Coding In-Depth Semistructured Interviews: Problems of Unitization and Intercoder Reliability and Agreement," *Sociological Methods and Research*, Vol. 42, No. 3, 2013, pp. 294–320.

Cassa, Christopher A., Rumi Chunara, Kenneth Mandl, and John S. Brownstein, "Twitter as a Sentinel in Emergency Situations: Lessons from the Boston Marathon Explosions," *PLOS Currents*, 2013.

Center for Army Lessons Learned, *Theater Security Cooperation: The Military Engagement Team, Lessons and Best Practices*, 2017.

Cervone, Guido, Elena Sava, Qunying Huang, Emily Schnebele, Jeff Harrison, and Nigel Waters, "Using Twitter for Tasking Remote-Sensing Data Collection and Damage Assessment: 2013 Boulder Flood Case Study," *International Journal of Remote Sensing*, Vol. 37, No. 1, 2016, pp. 100–124.

Chen, Hao, Maria Vasardani, and Stephan Winter, "Georeferencing Places from Collective Human Descriptions Using Place Graphs," *Journal of Spatial Information Science*, Vol. 17, 2018.

Cheng, Evelyn, "Just 10% of Trading is Regular Stock Picking, JP Morgan Estimates," *CNBC*, June 13, 2017. As of September 13, 2018:
https://www.cnbc.com/2017/06/13/death-of-the-human-investor-just-10-percent-of-trading-is-regular-stock-picking-jpmorgan-estimates.html

Chu, Zi, Steven Gianvecchio, Haining Wang, and Sushil Jajodia, "Detecting Automation of Twitter Accounts: Are You a Human, Bot, or Cyborg?" *IEEE Transactions on Dependable and Secure Computing*, Vol. 9, No. 6, 2012, pp. 811–824.

Ciere, Yvette, Debbie Jaarsma, Annemieke Visser, Robbert Sanderman, Evelien Snippe, and Joke Fleer, "Studying Learning in the Healthcare Setting: The Potential of Quantitative Diary Methods," *Perspectives in Medical Education*, Vol. 4, No. 4, 2015.

Collingwood, Loren, and John Wilkerson, "Tradeoffs in Accuracy and Efficiency in Supervised Learning Methods," *JITP 2011: The Future of Computational Social Science*, 2011.

Collins, Amy, "Your Questions Answered: Instagram API Changes," *Brandwatch*, April 3, 2018. As of November 18, 2019:
https://www.brandwatch.com/blog/your-questions-answered-instagram-api-changes/

Connable, Ben, *Embracing the Fog of War: Assessment and Metrics in Counterinsurgency*, Santa Monica, Calif.: RAND Corporation, MG-1086-DOD, 2012. As of November 7, 2019:
https://www.rand.org/pubs/monographs/MG1086.html

Connable, Ben, Michael J. McNerney, William Marcellino, Aaron Frank, Henry Hargrove, Marek N. Posard, S. Rebecca Zimmerman, Natasha Lander, Jasen J. Castillo, and James Sladden, *Will to Fight: Analyzing, Modeling, and Simulating the Will to Fight of Military Units*, Santa Monica, Calif.: RAND Corporation, RR-2341-A, 2018. As of November 7, 2019: https://www.rand.org/pubs/research_reports/RR2341.html

Constine, Josh, "Facebook Restricts APIs, Axes Old Instagram Platform Amidst Scandals," *TechCrunch*, April 4, 2018. As of October 2, 2018: https://techcrunch.com/2018/04/04/facebook-instagram-api-shut-down/

Cooner, Austin J., Yang Shao, and James B. Campbell, "Detection of Urban Damage Using Remote Sensing and Machine Learning Algorithms: Revisiting the 2010 Haiti Earthquake," *Remote Sensing*, Vol. 8, No. 10, 2016.

Coppersmith, Glen, Ryan Leary, Patrick Crutchley, and Alex Fine, "Natural Language Processing of Social Media as Screening for Suicide Risk," *Biomedical Informatics Insights*, Vol. 10, 2018.

Croicu, Mihai, and Ralph Sundberg, "UCDP Georeferenced Event Dataset Codebook Version 18.1," Department of Peace and Conflict Research, Uppsala University, 2018.

De Aenille, Conrad, "A.I. Has Arrived in Investing. Humans Are Still Dominating," *New York Times*, January 12, 2018.

Defense Advanced Research Projects Agency, "Visual Media Reasoning (VMR) (Archived)," webpage, undated. As of November 7, 2019: https://www.darpa.mil/program/visual-media-reasoning

Department of Defense Instruction 5132.14, *Assessment, Monitoring, and Evaluation Policy for the Security Cooperation Enterprise*, Washington, D.C.: Office of the Under Secretary of Defense for Policy, January 13, 2017.

Deployable Training Division, *Insights and Best Practices Focus Paper: Design and Planning*, Suffolk, Va.: Joint Staff J7, July 2013. As of January 3, 2020: http://www.jcs.mil/Portals/36/Documents/Doctrine/fp/design_and_planning_fp.pdf

Deshpande, Bala, "When Do Support Vector Machines Trump Other Classification Methods," Simafore, blog post, January 28, 2013. As of January 20, 2020, archived at https://web.archive.org/web/20190703071316/http://www.simafore.com/blog/bid/112816/When-do-support-vector-machines-trump-other-classification-methods

Domingos, Pedro M., "A Few Useful Things to Know About Machine Learning," *Communications of the ACM*, Vol. 55, No. 10, 2012, pp. 78–87.

Duggan, Maeve, and Joanna Brenner, "The Demographics of Social Media Users—2012," *Pew Research Center's Internet and American Life Project*, 2013.

Easter, Jonathan R., and Benon M. Hatangimana, *Bending the Spear: The Campaign Against the Lord's Resistance Army*, Monterey, Calif.: Naval Postgraduate School, December 2017.

Egel, Daniel, and Peter Glick, *Assessing the Stabilizing Effects of Development Programming: A Review of USAID's Evaluation Efforts in Afghanistan*, Santa Monica, Calif.: RAND Corporation, WR-1106-MSI, 2016. As of November 7, 2019: https://www.rand.org/pubs/working_papers/WR1106.html

Egel, Daniel, Charles P. Ries, Ben Connable, Todd C. Helmus, Eric Robinson, Isaac Baruffi, Melissa A. Bradley, Kurt Card, Kathleen Loa, Sean Mann, Fernando Sedano, Stephan B. Seabrook, and Robert Stewart, *Investing in the Fight: Assessing the Use of the Commander's Emergency Response Program in Afghanistan*, Santa Monica, Calif.: RAND Corporation, RR-1508-OSD, 2016. As of November 7, 2019:
https://www.rand.org/pubs/research_reports/RR1508.html

El-Kilany, Ayman, Amr Azzam, and Samhaa R El-Beltagy, "Using Deep Neural Networks for Extracting Sentiment Targets in Arabic Tweets," in Khaled Shaalan, Aboul Ella Hassanien, and Mohammed F. Tolba, eds., *Intelligent Natural Language Processing: Trends and Applications*, New York: Springer, 2018, pp. 3–15.

Federal Bureau of Investigation, Directorate of Intelligence, "FBI Intelligence Information Report (IIR) Dissemination Systems (FIDS)," undated. As of March 26, 2019:
https://www.archives.gov/files/records-mgmt/rcs/schedules/departments/department-of-justice/rg-0065/n1-065-10-025_sf115.pdf

Fiesler, Casey, Michaelanne Dye, Jessica L. Feuston, Chaya Hiruncharoenvate, Clayton J. Hutto, Shannon Morrison, Parisa Khanipour Roshan, Umashanthi Pavalanathan, Amy S. Bruckman, and Munmun De Choudhury, "What (or Who) Is Public? Privacy Settings and Social Media Content Sharing," *Proceedings of the 2017 ACM Conference on Computer Supported Cooperative Work and Social Computing*, 2017.

Forest, James J. F., *U.S. Military Deployments to Africa: Lessons from the Hunt for Joseph Kony and the Lord's Resistance Army*, MacDill Air Force Base, Fla.: Joint Special Operations University, Report 14-4, 2014.

Freedman, David, "From Association to Causation: Some Remarks on the History of Statistics," *Statistical Science*, Vol. 14, No. 3, August 1999.

Geitgey, Adam, "Machine Learning Is Fun Part 8: How to Intentionally Trick Neural Networks: A Look into the Future of Hacking," *Medium*, August 16, 2017. As of November 19, 2019:
https://medium.com/@ageitgey/machine-learning-is-fun-part-8-how-to-intentionally-trick-neural-networks-b55da32b7196

Gröger, Andre, Jonathan Hersh, Andrea Matranga, and Hannes Müller, "Analyzing Conflict from Space: Identification of Physical Destruction During the Syrian Civil War," unpublished research, 2018.

Haass, Felix, and Nadine Ansorg, "Better Peacekeepers, Better Protection? Troop Quality of United Nations Peace Operations and Violence Against Civilians," *Journal of Peace Research*, Vol. 55, No. 6, 2018.

Henderson, J. Vernon, Adam Storeygard, and David N. Weil, "Measuring Economic Growth from Outer Space," *American Economic Review*, Vol. 102, No. 2, 2012.

Hernandez, Alejandro S., Julian Ouellet, and Christopher J. Nannini, "Circular Logic and Constant Progress: IW Assessments in Afghanistan," in Leo J. Blanken, Hy Rothstein, and Jason J. Lepore, eds., *Assessing War: The Challenge of Measuring Success and Failure*," Washington, D.C.: Georgetown University Press, 2015.

Hoffman, Francis G., "Strategic Assessment and Adaptation: Reassessing the Afghanistan Surge Decision," *Naval War College Review*, Vol. 69, No. 3, 2016, p. 2. As of September 4, 2018:
http://digital-commons.usnwc.edu/cgi/viewcontent.cgi?article=1161&context=nwc-review

Hopkins, Daniel J., Eunji Kim, and Soojong Kim, "Does Newspaper Coverage Influence or Reflect Public Perceptions of the Economy?" *Research and Politics*, Vol. 4, No. 4, 2017.

Horne, Benjamin D., and Sibel Adali, "An Exploration of Verbatim Content Republishing by News Producers," *arXiv*, 2018.

Horns, Joshua J., Frederick R. Adler, Çağan H. Şekercioğlu, "Using Opportunistic Citizen Science Data to Estimate Avian Population Trends," *Biological Conservation*, Vol. 221, May 2018, pp. 151–159.

Jean, Neal, Marshall Burke, Michael Xie, W. Matthew Davis, David B. Lobell, and Stefano Ermon, "Combining Satellite Imagery and Machine Learning to Predict Poverty," *Science*, Vol. 353, No. 6301, August 19, 2016.

Johnston, Patrick B., and Anoop K. Sarbahi, "The Impact of US Drone Strikes on Terrorism in Pakistan," *International Studies Quarterly*, Vol. 60, No. 2, June 1, 2016.

Joint Doctrine Note 1-15, *Operation Assessment*, Washington, D.C.: Joint Chiefs of Staff, January 15, 2015.

Joint Publication 5-0, *Joint Planning*, Washington, D.C.: Joint Chiefs of Staff, June 16, 2017.

Joint Staff, *Commander's Handbook for Assessment Planning and Execution*, Suffolk, Va., September 9, 2011.

Jurgens, David, "That's What Friends Are for: Inferring Location in Online Social Media Platforms Based on Social Relationships," *Proceedings of the 7th International Conference on Weblogs and Social Media*, 2013.

Keller, Jared, "How the Army Decimated Joseph Kony's Messianic Death Cult Without Firing a Shot," *Task and Purpose*, October 19, 2017. As of May 6, 2019:
https://taskandpurpose.com/kony-uganda-army-psyops-report/

Koepfler, Jes A., and Kenneth R. Fleischmann, "Studying the Values of Hard-to-reach Populations: Content Analysis of Tweets by the 21st Century Homeless," *Proceedings of the 2012 iConference: Association for Computing Machinery*, 2012.

Kohavi, Ron, "A Study of Cross-Validation and Bootstrap for Accuracy Estimation and Model Selection," *International Joint Conference on Artificial Intelligence*, 1995.

Krippendorff, Klaus, *Content Analysis: An Introduction to Its Methodology*, Newbury Park, Calif.: SAGE Publications, Inc., 2004.

Küçük, Dilek, and Fazli Can, "A Tweet Dataset Annotated for Named Entity Recognition and Stance Detection," *arXiv*, January 16, 2019.

LaRivee, Dave, "Best Practices Guide for Conducting Assessments in Counterinsurgencies," *Small Wars Journal*, 2011.

Lozhnikov, Nikita, Leon Derczynski, and Manuel Mazzara, "Stance Prediction for Russian: Data and Analysis," *International Conference in Software Engineering for Defence Applications*, 2018.

Marcum, Richard A., Curt H. Davis, Grant J. Scott, and Tyler W. Nivin, "Rapid Broad Area Search and Detection of Chinese Surface-to-Air Missile Sites Using Deep Convolutional Neural Networks," *Journal of Applied Remote Sensing*, Vol. 11, No. 4, 2017.

Maxwell, Aaron E., Timothy A. Warner, and Fang Fang, "Implementation of Machine-Learning Classification in Remote Sensing: An Applied Review," *International Journal of Remote Sensing*, Vol. 39, No. 9, 2018.

McCormick, Tyler H., Hedwig Lee, Nina Cesare, Ali Shojaie, and Emma S. Spiro, "Using Twitter for Demographic and Social Science Research: Tools for Data Collection and Processing," *Sociological Methods & Research*, Vol. 46, No. 3, August 2017.

McCready, Daniel, "Developing Remotely piloted Air-to-Air Combat Capabilities," Unmanned Systems Summit, Alexandria, Va., March 14, 2019.

McKenzie, David, "How Can Machine Learning and Artificial Intelligence Be Used in Development Interventions and Impact Evaluations?" *World Bank Blog*, March 5, 2018. As of September 5, 2018: https://blogs.worldbank.org/impactevaluations/how-can-machine-learning-and-artificial-intelligence-be-used-development-interventions-and-impact

Mellander, Charlotta, José Lobo, Kevin Stolarick, and Zara Matheson, "Night-Time Light Data: A Good Proxy Measure for Economic Activity?" *PLOS One*, Vol. 10, No. 10, 2015.

Mirończuk, Marcin Michał, and Jarosław Protasiewicz, "A Recent Overview of the State-of-the-Art Elements of Text Classification," *Expert Systems with Applications*, Vol. 106, 2018, pp. 36–54.

Mueller, Hannes, and Christopher Rauh, "Reading Between the Lines: Prediction of Political Violence Using Newspaper Text," *Empirical Studies of Conflict*, Working Paper No. 2, 2016.

Mushen, Emily and Jonathan Schroden, *Are We Winning? A Brief History of Military Operations Assessment*, Washington, D.C.: CNA, September 2014.

North Atlantic Treaty Organization, *Allied Joint Doctrine for Countering—Improvised Explosive Devices*, Allied Joint Publication 3.15(A), March 2011. As of July 17, 2018: http://www.dtra.mil/Portals/61/Documents/Missions/NATO%20AJP-3.15(A)%20ALLIED%20C-IED%20MAR%2011.pdf?ver=2017-03-10-134619-480

O'Halloran, Kay L., Sabine Tan, Peter Wignell, John A. Bateman, Duc-Son Pham, Michele Grossman, and Andrew Vande Moere, "Interpreting Text and Image Relations in Violent Extremist Discourse: A Mixed Methods Approach for Big Data Analytics," *Terrorism and Political Violence*, Vol. 31, No. 3, 2016.

Pan, Da, Meishan Zhang, and Guohong Fu, "Chinese Sentiment Analysis Exploiting Heterogeneous Segmentations," *CCL*, 2016, pp. 387–398.

Parthasarathy, Ramya, Vijayendra Rao, and Nethra Palaniswamy, "Deliberative Inequality: A *Text-as-Data* Study of Tamil Nadu's Village Assemblies," *World Bank*, Policy Research Working Paper 8119, June 2017. As of September 6, 2018: http://documents.worldbank.org/curated/en/582551498568606865/pdf/WPS8119.pdf

Paruchabutr, Gittipong, Christopher Parapone, Geoffrey Demarest, Grant Martin, Christopher Kirk, Cody Brown, Curtis Clayton, William Harris, Thomas Heilman, and Brian Kaiser, *SOF Campaign Planner's Handbook of Operational Art and Design*, Fort Bragg, N.C.: U.S. Army Special Operations Command, September 16, 2014.

Patra, Braja Gopal, Dipankar Das, and Amitava Das, "Sentiment Analysis of Code-Mixed Indian Languages: An Overview of SAIL_Code-Mixed Shared Task@ ICON-2017," *arXiv*, March 18, 2018.

Paul, Amy, Craig Jolley, and Aubra Anthony, *Reflecting the Past, Shaping the Future: Making AI Work for International Development*, Washington, D.C.: U.S. Agency for International Development, 2018.

Pearl, Judea, "The Seven Tools of Causal Inference, with Reflections on Machine Learning," *Communications of Association for Computing Machinery*, 2018.

Petit, Brian, "The Fight for the Village: Southern Afghanistan," *Military Review*, May–June 2011.

Pierskalla, Jan H., and Florian M. Hollenbach, "Technology and Collective Action: The Effect of Cell Phone Coverage on Political Violence in Africa," *American Political Science Review*, Vol. 107, No. 2, 2013, pp. 207–224.

Pottinger, Matt, Michael T. Flynn, and Paul D. Batchelor, "Fixing Intel: A Blueprint for Making Intelligence Relevant in Afghanistan," Center for a New American Security, January 2010. As of October 1, 2018:
https://www.cnas.org/publications/reports/fixing-intel-a-blueprint-for-making-intelligence-relevant

Pritt, Mark, and Gary Chern, "Satellite Image Classification with Deep Learning," *2017 IEEE Applied Imagery Pattern Recognition Workshop*, 2017.

Public Law 117-172, Lord's Resistance Army Disarmament and Northern Uganda Recovery Act of 2009, May 24, 2010.

Public Law 114-328, National Defense Authorization Act for Fiscal Year 2017, December 23, 2016.

Raleigh, Clionadh, and Caitriona Dowd, "Armed Conflict Location and Event Data Project (ACLED) Codebook," Armed Conflict Location and Event Data Project, 2015. As of October 19, 2018:
https://www.acleddata.com/wp-content/uploads/2015/01/ACLED_Codebook_2015.pdf

Robinson, Eric, Daniel Egel, Patrick B. Johnston, Sean Mann, Alexander D. Rothenberg, and David Stebbins, *When the Islamic State Comes to Town: The Economic Impact of Islamic State Governance in Iraq and Syria*, Santa Monica, Calif.: RAND Corporation, RR-1970-RC, 2017. As of November 7, 2019:
https://www.rand.org/pubs/research_reports/RR1970.html

Robinson, Linda, Daniel Egel, and Ryan Andrew Brown, *Measuring the Effectiveness of Special Operations*, Santa Monica, Calif.: RAND Corporation, RR-2504-A, 2019.

Rosa, Mark, and Jan Feyereisl, "A Framework for Searching for General Artificial Intelligence," *arXiv*, 2016.

Ross, Casey, and Ike Swetlitz, "IBM Pitched Its Watson Supercomputer as a Revolution in Cancer Care. It's Nowhere Close," *STAT*, September 5, 2017. As of September 13, 2018:
https://www.statnews.com/2017/09/05/watson-ibm-cancer/

Ryman-Tubb, Nick F., Paul Krause, and Wolfgang Garn, "How Artificial Intelligence and Machine Learning Research Impacts Payment Card Fraud Detection: A Survey and Industry Benchmark," *Engineering Applications of Artificial Intelligence*, Vol. 76, 2018, pp. 130–157.

Schroden, Jonathan, "Why Operations Assessments Fail: It's Not Just the Metrics," *Naval War College Review*, Vol. 64, No. 4, Autumn 2011.

Schroden, Jonathan, Rebecca Thomasson, Randy Foster, Mark Lukens, and Richard Bell, "A New Paradigm for Assessment in Counterinsurgency," *Military Operations Research*, Vol. 18, No. 3, 2013, pp. 5–20.

Signal Program on Human Security and Technology, *Satellite Imagery Interpretation Guide: Intentional Burning of Tukuls*, Harvard Humanitarian Initiative, 2015.

Sloan, Luke, Jeffrey Morgan, Pete Burnap, and Matthew Williams, "Who Tweets? Deriving the Demographic Characteristics of Age, Occupation and Social Class from Twitter User Meta-Data," *PLOS One*, Vol. 10, No. 3, 2015.

Sotire, Robert B., *Measuring Performance and Effectiveness in Irregular Warfare: Preventing Dysfunctional Behavior*, Newport, R.I.: Naval War College, 2009.

Sundberg, Ralph, and Erik Melander, "Introducing the UCDP Georeferenced Event Dataset," *Journal of Peace Research*, Vol. 50, No. 4, 2013, pp. 523–532.

Tao, Ran, Daniel Strandow, Michael Findley, Jean-Claude Thill, and James Walsh, "A Hybrid Approach to Modeling Territorial Control in Violent Armed Conflicts," *Transactions in GIS*, Vol. 20, No. 3, 2016, pp. 413–425.

Tornes, Adam, "Introducing the Premium Full-Archive Search Endpoint," Twitter Developer Blog, February 1, 2018. As of November 7, 2019:
https://blog.twitter.com/developer/en_us/topics/tools/2018/access-the-full-history-of-tweets.html#

Tucker, Patrick, "The Apps They Carried: Software, Big Data, and the Fight for Mosul," *DefenseOne*, October 17, 2016. As of October 3, 2018:
https://www.defenseone.com/technology/2016/10/apps-they-carried-software-bi
g-data-and-fight-mosul/132409/

Turney, Peter D., "Thumbs Up or Thumbs Down? Semantic Orientation Applied to Unsupervised Classification of Reviews," *Proceedings of the Association for Computational Linguistics*, 2002.

Upshur, William P., Jonathan W. Roginski, and David J. Kilcullen, "Recognizing Systems in Afghanistan: Lessons Learned and New Approaches to Operational Assessments," *Prism*, Vol. 3, No. 3, June 2012.

U.S. Africa Command, *U.S. Forces Transition Counter-LRA Mission to Broader Security and Stability Activities*, March 29, 2017.

U.S. Air Force, *Annex 2-0: Global Integrated Intelligence, Surveillance & Reconnaissance Operations*, Air University, Curtis E. Lemay Center for Doctrine Development and Education, 2015. As of October 1, 2018:
https://www.doctrine.af.mil/Portals/61/documents/Annex_2-0/2-0-D14-ISR-Intel-Disciplines.pdf

U.S. Army War College, *Campaign Planning Handbook*, Carlisle Barracks, Pa.: Department of Military Strategy, Planning, and Operations, 2016.

U.S. House of Representatives, *The Joint Improvised Explosive Device Defeat Organization: DOD's Fight Against IEDs Today and Tomorrow*, Committee on Armed Services, Subcommittee on Oversight & Investigations November 2008. As of July 17, 2018:
https://democrats-armedservices.house.gov/_cache/files/c/f/cfddccb2-fc15-4a3d-b7e3-50fe3ea68eca/D09F0BEF55D1B39D2CC196408918781D.jieddo-report-11-08-vf.pdf

Wang, Alex Hai, "Detecting Spam Bots in Online Social Networking Sites: A Machine Learning Approach," *24th Annual IFIP WG 11.3 Working Conference on Data and Applications Security and Privacy (DBSEC)*, Rome, Italy, June 2010, pp. 335–342.

Weidmann, Nils B., "A Closer Look at Reporting Bias in Conflict Event Data," *American Journal of Political Science*, Vol. 60, No. 1, January 2016.

The White House, "Letter from the President to the Speaker of the House of Representatives and the President Pro Tempore of the Senate Regarding the Lord's Resistance Army," October 14, 2011.

Yang, Chenxi, Wenjing Li, and Zhiyong Lin, "Vehicle Object Detection in Remote Sensing Imagery Based on Multi-Perspective Convolutional Neural Network," *International Journal of Geo-Information*, Vol. 7. No. 7, 2018.

YouTube, "YouTube-8M Segments Dataset," webpage, undated. As of November 7, 2019:
https://research.google.com/youtube8m/

Yu, Fisher, Ari Seff, Yinda Zhang, Shuran Song, Thomas Funkhouser, and Jianxiong Xiao, "LSUN: Construction of a Large-Scale Image Dataset using Deep Learning with Humans in the Loop," *arXiv*, June 4, 2016.

Zaydman, Mikhail, *Tweeting About Mental Health: Big Data Text Analysis of Twitter for Public Policy*, Santa Monica, Calif.: RAND Corporation, RGSD-391, 2017. As of November 7, 2019: https://www.rand.org/pubs/rgs_dissertations/RGSD391.html

Zhao, Bo, and Daniel Z. Sui, "True Lies in Geospatial Big Data: Detecting Location Spoofing in Social Media," *Annals of GIS*, Vol. 23, No. 1, 2017, pp. 1–14.